D0939317

# SYSTEMC: FROM THE GROUND UP

# SYSTEMC: FROM THE GROUND UP

By

**David C. Black** and **Jack Donovan**
*Eklectic Ally, Inc.*

 Springer

David C. Black
Jack Donovan
Eklectic Ally, Inc.

Black, David C. (David Christopher), 1956-
    SystemC: from the ground up / by David C. Black and Jack Donovan
        p. cm.
    Includes index.
    ISBN 1-4020-7988-5        eBook ISBN 1-4020-7989-3
    1. System design. 2. C++ (Computer program language). I. Donovan, Jack, 1956-- II. Title

QA76.9.S88B578 2004
005.13í3--dc22

                                                                    2004042112


ISBN 0-387-29240-3 (SC)        e-ISBN 1-4020-7989-3        Printed on acid-free paper.
ISBN 978-0387-29240-3
ISBN 1-4020-7988-5 (HC)

First softcover printing 2006
©2004 Springer Science+Business Media, Inc. (hardcover edition)
All rights reserved. This work may not be translated or copied in whole or in part without the written permission of the publisher (Springer Science+Business Media, Inc., 233 Spring Street, New York, NY 10013, USA), except for brief excerpts in connection with reviews or scholarly analysis. Use in connection with any form of information storage and retrieval, electronic adaptation, computer software, or by similar or dissimilar methodology now know or hereafter developed is forbidden.
The use in this publication of trade names, trademarks, service marks and similar terms, even if the are not identified as such, is not to be taken as an expression of opinion as to whether or not they are subject to proprietary rights.

Printed in the United States of America.

9 8 7 6 5 4 3 2 1        SPIN  11564805

springeronline.com

# Dedication

*This book is dedicated to our spouses*
*Pamela Black and Carol Donovan,*
*and to our children*
*Christina, Loretta, & William Black,*
*Chris, Karen, Jenny, & Becca Donovan*

# Contents

# Preface

## Why this Book

The first question any reader should ask is "Why this book?" We decided to write this book after learning SystemC and after using minimal documents to help us through the quest of becoming comfortable with the language's finer points. After teaching several SystemC classes, we were even more convinced that an introductory book focused on the SystemC language was needed. We decided to contribute such a book.

This book is about SystemC. It focuses on enabling the reader to master the language. The reader will be introduced to the syntax and structure of the language, and the reader will also learn about the features and usage of SystemC that makes it a tool to shorten the development cycle of large system designs.

We allude to system design techniques and methods by way of examples throughout the book. Several books that discuss system-level design methodology are available, and we believe that SystemC is ideally suited to implement many of these methods. After reading this resource, the reader should not only be adept at using SystemC constructs efficiently, but also have an appreciation of how the constructs work together and how they can be used to create high performance simulation models.

We believe there is enough information to convey about the SystemC language to justify this stand-alone book. We hope you agree. We also believe that there is enough material for a second book that focuses on using SystemC to implement these system-level design methods. With reader encouragement, the authors hope to start on a second book that delves deeper into the application of the language (after recovering from the writing of this book).

## Prerequisites for this Book

As with every technical book, the authors must write the content assuming a basic level of understanding; this assumption avoids repeating most of an engineering undergraduate curriculum. For this book, we assumed that the reader has a working knowledge of C++ and minimal knowledge of hardware design.

For C++ skills, we do not assume that the reader is a "wizard". Instead, we assumed that you have a good knowledge of the syntax, the object-oriented features, and the methods of using C++. The authors have found that this level of C++ knowledge is universal to current or recent graduates with a computer science or engineering degree from a four-year university.

Interestingly, the authors have also found that this level of knowledge is lacking for most ASIC designers with 10 or more years of experience. For those readers, assimilating this content will be quite a challenge but not an impossible one.

For readers without any understanding of C++ or for those who may be rusty, we recommend finding a good C++ class at a community college or taking advantage of many of the online tutorials. For a list of sources, see Chapter 15. We find (from our own experience) that those who have learned several procedural languages (like FORTRAN or PL/I) greatly underestimate the difficulty of learning a modern object-oriented language.

To understand the examples completely, the reader will need minimal understanding of digital electronics.

## Book Conventions

Throughout this book, we include many syntax and code examples. We've chosen to use an example-based approach because most engineers have an easier time understanding examples rather than strict Backus-Naur Form[1] (BNF) syntax or precise library declarations. Syntax examples illustrate the code in the manner it may be seen in real use with placeholders for user-specified items. For more complete library information, we refer the reader to the SystemC Language Reference Manual (LRM) currently at version 2.0.1, which you can download from www.systemc.org.

Code will appear in `monotype Courier` font. Keywords for both C/C++ and SystemC will be shown in **bold**. User-selectable syntax items are in *italics* for emphasis. Repeated items may be indicated with an ellipsis (...) or subscripts. The following is an example:

```
wait(name.posedge_event() | event_i...);
if (name.posedge()) { //previous delta-cycle
   //ACTIONS...
```

*Figure 1.* Example of Sample Code

In addition, the following are standard graphical notations. The terminology will be presented as the book progresses.

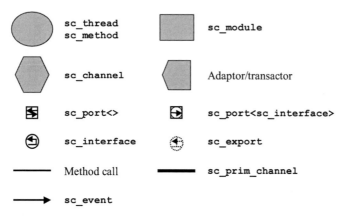

*Figure 2.* Standard Graphical Notations

---

[1] John Backus and Peter Naur first introduced BNF as a formal notation to describe the syntax of a given language. NAUR, Peter (ed.), "Revised Report on the Algorithmic Language ALGOL 60.", Communications of the ACM, Vol. 3 No.5, pp. 299-314, May 1960

SystemC uses a naming convention where most SystemC specific identifiers are prefixed with **sc_** or **SC_**. This convention is reserved for the SystemC library, and you should not use it in end-user code (your code).

## About the Examples

To introduce the syntax of SystemC and demonstrate its usage, we have filled this book with examples. Most examples are not real-world examples. Real examples just become too cluttered too fast. The goal of these examples is to communicate concepts clearly; we hope that the reader can extend them into the real world. For the most part, we used a common theme of an automobile for the examples.

By convention, we show syntax examples stylistically as if SystemC is a special language, which it is not. We hope that this presentation style will help you apply SystemC on your first coding efforts. If you are looking for the C++ declarations, please browse the LRM or look directly into the SystemC Open SystemC Initiative reference source code (www.systemc.org).

## How to Use this Book

The authors designed this book primarily for the student or engineer new to SystemC. This book's structure is best appreciated by reading sequentially from beginning to end. A reader already familiar with SystemC will find this content to be a great reference.

Chapters 1 through 3 provide introductions and overviews of the language and its usage. Methodology is briefly discussed.

For the student or engineer new to SystemC, the authors present the language from the bottom up and explain the topics from a context of C++ with ties to hardware concepts. We provide exercises at the end of Chapters 4 through 14 to reinforce the concepts presented in the text. Chapter 15 backs up the basic language concepts with a discussion of areas to watch when designing, writing, or using SystemC in a production environment.

For the student or engineer already familiar with SystemC, Chapters 4 through 11 will provide some interesting background and insights into the language. You can either skip or read these early chapters lightly to pick up more nuances of the language. The content here is not meant to be a complete description of the language. For a thorough description, the reader is referred to the SystemC LRM. Chapters 12 through 14 provide intermediate to advanced material.

For the advanced reader, Chapter 15 provides performance tips and information about gotchas and tidbits that may help with some of the day-to-day usage of SystemC.

For the instructor, this book may provide part of an advanced undergraduate class on simulation or augment a class on systems design.

In most of the examples presented in the book, the authors show code fragments only so as to emphasize the points being made. Examples are designed to illustrate specific concepts, and as such are toy examples to simplify learning. Complete source code for all examples and exercises is available for download from www.EklecticAlly.com/Book as a compressed archive. You will need this book to make best use of these files.

## SystemC Background

SystemC is a system design language based on C++. As with most design languages, SystemC has evolved. Many times a brief overview of the history of language will help answer the question "Why do it that way?" We include a brief history of SystemC and the Open SystemC Initiative to help answer these questions.

## The Evolution of SystemC

SystemC is the result of the evolution of many concepts in the research and commercial EDA communities. Many research groups and EDA companies have contributed to the language. A timeline of SystemC is included below.

Table 1-1. Timeline of Development of SystemC

| Date | Version | Notes |
|------|---------|-------|
| Sept 1999 | 0.9 | First version; cycle-based |
| Feb 2000 | 0.91 | Bug fixes |
| Mar2000 | 1.0 | Widely accessed major release |
| Oct 2000 | 1.0.1 | Bug fixes |
| Feb 2001 | 1.2 | Various improvements |
| Aug 2002 | 2.0 | Add channels & events; cleaner syntax |
| Apr 2002 | 2.0.1 | Bug fixes; widely used |
| June 2003 | 2.0.1 | LRM in review |
| Spring 2004 | 2.1 | LRM submitted for IEEE standard |

SystemC started out as a very restrictive cycle-based simulator and "yet another" RTL language. The language has evolved (or is evolving) to a true system design language that includes both software and hardware concepts. Although SystemC does not specifically support analog hardware or mechanical components, there is no reason why these aspects of a system cannot be modeled with SystemC constructs or with co-simulation techniques.

# Open SystemC Initiative

Several of the organizations that contributed heavily to the language development efforts realized very early that any new design language must be open to the community and not be proprietary. As a result, the Open SystemC Initiative (OSCI) was formed in 1999. OSCI was formed to:

- Evolve and standardize the language
- Facilitate communication among the language users and tool vendors
- Enable adoption
- Provide the mechanics for open source development and maintenance

### The SystemC Verification Library

As you will learn while reading this book, SystemC consists of the language and potential methodology-specific libraries. The authors view the SystemC Verification (SCV) library as the most significant of these libraries. This library adds support for modern high-level verification language concepts such as constrained randomization, introspection, and transaction recording. The first release of the SCV library occurred in December of 2003 after over a year of Beta testing.

### Current Activities with OSCI

At present, the Open SystemC Initiative is involved with completing the LRM and submitting it to the Institute of Electrical and Electronics Engineers (IEEE) for standardization. Additionally, sub-committees are studying such topics as synthesis subsets and formalizing terminology concerning levels of abstraction.

Chapter 1

# AN OVERVIEW TO SYSTEM DESIGN USING SYSTEMC

## 1.1 Introduction

SystemC is a system design language that has evolved in response to a pervasive need for a language that improves overall productivity for designers of electronic systems. Typically, today's systems contain application-specific hardware and software. Furthermore, the hardware and software are usually co-developed on a tight schedule, the systems have tight real-time performance constraints, and thorough functional verification is required to avoid expensive and sometimes catastrophic failures.

SystemC offers real productivity gains by letting engineers design both the hardware and software components together as these components would exist on the final system, but at a high level of abstraction. This higher level of abstraction gives the design team a fundamental understanding early in the design process of the intricacies and interactions of the entire system and enables better system trade offs, better and earlier verification, and over all productivity gains through reuse of early system models as executable specifications.

## 1.2   Language Comparison

Strictly speaking, SystemC is not a language, but rather a class library within a well established language, C++. SystemC is not a panacea that will solve every design productivity issue. However, when SystemC is coupled with the SystemC Verification Library, it does provide in one language many of the characteristics relevant to system design and modeling tasks that are missing or scattered among the other languages. Additionally, SystemC provides a common language for software and hardware, C++.

Several languages have emerged to address the various aspects of system design. Although Ada and Java have proven their value, C/C++ is predominately used today for embedded system software. The hardware description languages (HDLs), VHDL and Verilog, are used for simulating and synthesizing digital circuits. Vera and *e* are the languages of choice for functional verification of complex application-specific integrated circuits (ASICs). SystemVerilog is a new language that evolves the Verilog language to address many hardware-oriented system design issues. Matlab and several other tools and languages such as SPW and System Studio are widely used for capturing system requirements and developing signal processing algorithms.

Figure 1-1 highlights the application of these and other system design languages. Each language occasionally finds use outside its primary domain, as the overlaps in Figure 1-1 illustrate.

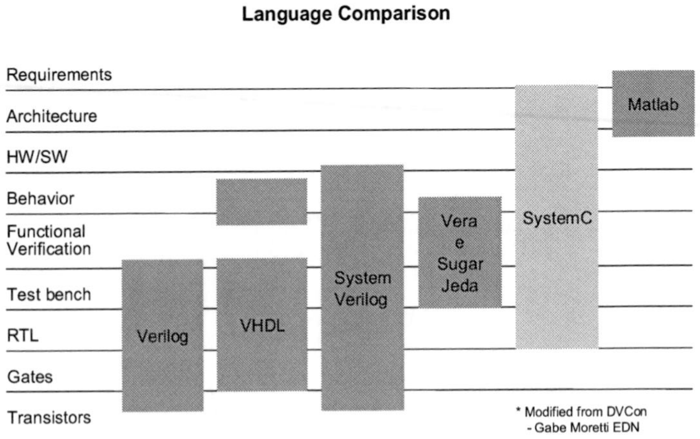

**Language Comparison**

*Figure 1-1*. SystemC contrasted with other design languages

## 1.3 Design Methods

Design methods surrounding SystemC are currently maturing and vary widely. In the next few years, these methods will likely settle into a cohesive design methodology (with a few variants among certain industry segments). The resulting methodology will feel similar to the methodologies in use today, but at higher levels of abstraction. To some, the concept of using one unified language for hardware and software development appears revolutionary, but this concept is clearly an evolutionary path for those who frequently work in both domains.

Although tools and language constructs exist in SystemC to support register-transfer-level (RTL) modeling and synthesis, a major reason for using the language is to work at higher abstraction levels than RTL. SystemC's ability to model RTL designs enables support of design blocks generated by higher level (behavioral or graphical entry) synthesis tools or to support legacy design blocks.

## 1.4 What's Next

Now that we have given you a map of where SystemC fits into the modern design language landscape, we will discuss some motivation for learning SystemC.

Unless a new language solves a problem that current languages do not address, you have no practical reason for learning that language (other than academic curiosity). The rest of this chapter will discuss the set of problems that you need to address with new methodologies and the types of solutions that SystemC is enabling. Understanding SystemC's solutions should help you understand some of the language trade offs that you may question when you dive into language details presented in Chapter 3, SystemC Overview.

## 1.5 Enhancing Productivity with SystemC

The primary motivation for using SystemC is to attain the productivity increases required to design modern electronic systems with their ever increasing complexity. Without productivity advances, many new system concepts will be impractical. In the next sections, we will examine system complexity, methods for attacking this complexity, and SystemC-enabled solutions.

### 1.5.1 Increasing Design Complexity

The primary driver leading to the need for a new system design language is the same that previously lead to the need for the current design languages: increasing design complexity.

Modern electronic systems consist of many sub-systems and components, but we will focus primarily on hardware, software, and algorithms. In modern systems, each of these disciplines has become more complex. Likewise, the interaction has become increasingly complex.

Interactions imply that trade offs between the domains are becoming more important for meeting customer requirements. System development teams find themselves asking questions like, "Should this function be implemented in hardware, software, or with a better algorithm?" Systems are so complex, just deriving specifications from customer requirements has become a daunting task.

Figure 1-2 illustrates the complexity issues for hardware design in a large system on a chip (SoC) design. The figure shows three sample designs from three generations: yesterday, today (2003), and tomorrow. In reality, tomorrow's systems are being designed today. The bars for each generation imply the code complexity for four common levels of abstraction associated with system hardware design:

- Architecture
- Behavioral
- RTL
- Gates

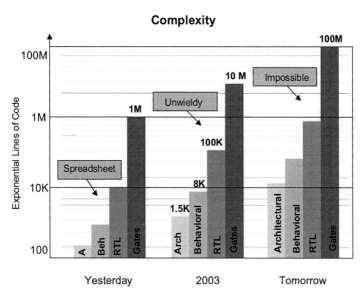

*Figure 1-2.* Design Complexity from Different Design Generation

Today's integrated circuits (ICs) often exceed 10 million gates, which conservatively translates to one hundred thousand lines of RTL code. Today's designs are practical because of the methodologies that apply RTL synthesis for automated generation of gates. Tomorrow's ICs, which are being designed today, will exceed one hundred million gates, or roughly one million lines of RTL code, if written using today's methodologies.

Notice that Figure 1-2 considers only a single integrated circuit. It does not reflect the greater complexity of a system with several large chips (ASIC or field-programmable gate array, FPGA) and gigabytes of application software. Many stop-gap approaches are being applied, but the requirement for a fundamentally new approach is clear.

### 1.5.2 Facing Design Complexity

SystemC supports several techniques for addressing the complexity of modern designs. Today's design community uses several approaches for attacking the complexity issues that come with complex system design:

- Abstraction
- Design reuse
- Team discipline
- Project reuse
- Automation

Let's look at each of these approaches in turn.

### 1.5.2.1  Abstraction

In the past, the primary technique for managing complexity for both the software and hardware community has been to raise the level of abstraction used for design. This approach will be the primary technique for the future as well.

For software developers, the transition was from assembler code to higher level languages like FORTAN and PL/I, and then to even more abstract languages like Lisp, Ada, and multi-paradigm languages such as C++. Today's software practitioners grapple with a plethora of fifth-generation languages as they strive to find the appropriate level of abstraction, expressiveness, flexibility, and performance required to develop and maintain modern systems. Modular programming, data hiding, object-oriented design, generic programming, exception and constraint handling all aim to manage complexity.

For the hardware community, the path started with polygon definitions hand drawn on Mylar film. Gate-level design was enabled through a variety of schematic capture tools and gate-level languages that appear very crude by today's standards. In the late 1980s, VHDL and Verilog enabled process-independent ASIC simulation, documentation, and net list development, and eventually RTL synthesis.

Unfortunately, the synthesizable RTL subsets of VHDL and Verilog force hardware designers to specify behavior on every clock edge, which offers less abstraction than C offers the software designer. While object-oriented software design and C++ accelerated in the 1990s, EDA vendors attempted to keep pace by offering behavioral synthesis tools that could build circuits from more abstract VHDL and Verilog descriptions. For a variety of reasons, designers gave these tools a cool reception, and major EDA vendors shifted their focus toward design reuse and meeting designer productivity requirements.

As a result, the hardware community has not coherently raised the design abstraction level beyond RTL for almost 20 years, and the community has instead focused on design reuse, team discipline, and automation.

### 1.5.2.2 Design Reuse

Reuse has emerged as the dominant productivity technique for RTL and software design. Many excellent books have been written on this subject[2]. Reuse will continue to be a major component of any new methodology to address increased complexity.

Platform-based design is an evolution of design reuse to higher levels of abstraction. Instead of reusing lower levels of design, whole compute platforms are now reused and programmed for specific applications[3].

Many times, we refer to design reuse as *external reuse* to distinguish this technique from *project reuse* discussed in a later section.

### 1.5.2.3 Team Discipline

When we refer to team discipline, we refer to the tools and techniques used to bring productivity to each engineer and the interactions among engineers. This area encompasses anything from revision control to written specifications and other documentation to design and code reviews.

In the past, team discipline techniques have been mostly applied locally and without coordination to the architecture, software, and hardware groups. Furthermore, team discipline has been applied with uneven acceptance among these groups. In a highly disciplined organization, the deliverables for each group may be defined, but this level of discipline is still not the norm. To face the complexity issues for next generation designs, team discipline needs to be more evenly applied across all of the system design disciplines, coordinated more closely, and augmented with more tools.

The Software Engineering Institute's (SEI) Capability Maturity Model (CMM) has contributed much to elevating the software development process from an art to a science. The hardware community has embraced this notion with the MORE (Measure of Reuse Excellence) grading system, but all

---

[2] Keating, M., Bricaud, P. 2002. *Reuse Methodology Manual for System-On-A-Chip Designs.* Norwell, Massachusetts: Kluwer Academic Publishers.

[3] Chang, H., Cooke, L., Hunt, M., Martin, G., McNelly, A., Todd, L. 1999. *Surviving the SOC Revolution: A Guide to Platform-Based Design.* Norwell, Massachusetts: Kluwer Academic Publishers.

system designers would do well to adopt similar philosophies across the entire spectrum of system design.

### 1.5.2.4 Project Reuse

Project reuse is the term we use to describe code that is defined for a project and is reused by others within the project or on the next project. Engineers on the project may reuse the code at the original level of abstraction, or they may refine the code to a new and lower level of abstraction by adding design details. Project reuse will allow code developed by the architecture group to be reused by the software group, hardware functional verification group, and the hardware design group. We will discuss the SystemC mechanism for project reuse, refinement with adaptors, in the next chapter and in greater detail in Chapter 13, Custom Channels.

### 1.5.2.5 Automation

Another important method for tackling design complexity is design process automation. The EDA industry has continually provided tools that improve designer productivity, and will continue to do so.

This productivity comes with a price—both monetary and team discipline. Most EDA tools come with coding guidelines to maximize tool performance and engineer productivity. In particular, with the exponentially increasing amount of RTL code required in modern systems, automatic code generation will play an increasing role, and adherence to guidelines for the code generation tools will be essential. Team discipline techniques go a long way towards optimal leveraging of the EDA tools they use.

In addition to automatic code generation, automation written by the development team (usually during unpaid overtime) has always been necessary to some extent, and it will continue to be of prime importance for system design flows.

### 1.5.3 SystemC and Methods to Attack Complexity

SystemC supports all of the complexity attacking methods just discussed (or else we would not have included them in a book about SystemC). SystemC efficiently supports all of these methods because it leverages C++.

One of SystemC's greatest strengths is its ability to provide higher levels of abstraction for all components of a design. Managing abstraction is the strongest weapon in combating complexity.

C++ programmers have been applying design reuse and enforcing team discipline for years by leveraging the features in C++ and other tools, and they have been distributing their software with high quality to a wide number of compute platforms. A great example is the relatively high quality of free (or nearly free) software available over the web via the GNU license and others.

SystemC implements project reuse by leveraging C++ interfaces to separate communications from algorithms. This leveraging allows independent refinement of a block's functionality and communication (I/O) through the use of interfaces and adapters as lightly described in the next chapter and discussed in detail in Chapter 13, Custom Channels.

Over two dozen EDA companies currently support SystemC by providing automation solutions. Again, the C++ community provides a plethora of design productivity tools that are cheaply or freely available over the Internet.

Facing design complexity effectively with these five techniques is why SystemC is emerging as today's design language standard and is adding users each day.

The next chapter explores a new system design methodology, a transaction level model-based methodology, enabled by SystemC.

# Chapter 2

# TLM-BASED METHODOLOGY

This chapter examines a methodology that enables you to model your large system designs at higher level of abstraction and realize actual productivity gains offered by SystemC.

## 2.1 Transaction-Level Modeling Overview

In the past, when many systems were a more manageable size, a system could be grasped by a single person known by a variety of titles such as system architect, chief engineer, lead engineer, or project engineer. This guru may have been a software engineer, hardware engineer, or algorithm expert depending on the primary technology leveraged for the system. The complexity was such that this person could keep most or all of the details in his or her head, and this technical leader was able to use spreadsheets and paper-based methods to communicate thoughts and concepts to the rest of the team.

The guru's background usually dictated his or her success in communicating requirements to each of the communities involved in the design of the system. The guru's past experiences also controlled the quality of the multi-discipline trade offs such as hardware implementation versus software implementation versus algorithm improvements.

In most cases, these trade offs resulted in conceptual disconnects among the three groups. For example, cellular telephone systems consist of very complex algorithms, software, and hardware, and teams working on them have traditionally leveraged more rigorous but still ad-hoc methods.

These methods usually consist of a software-based model; sometimes called a system architectural model (SAM), written in C, Java, or a similar language. The model is a communication vehicle between algorithm, hardware, and software groups. The model may be used for algorithmic refinement or used as basis for deriving hardware and software subsystem specifications. The exact parameters modeled are specific to the system type and application, but the model is typically un-timed (more on this topic in the following section). Typically, each team then uses a different language to refine the design for their portion of the system. The teams leave behind the original multi-discipline system model and in many cases, any informal communication among the groups.

With rapidly increasing design complexity and the rising cost of failure, system designers in most product domains will need a similar top-down approach but with an improved methodology. An emerging system design methodology based on Transaction-Level Modeling (TLM) is evolving from the large system design methodology discussed above. This emerging methodology has significantly more external and project design reuse enabled by a language like SystemC.

Transaction-level modeling is an emerging concept without precise definitions. A working group of Open SystemC Initiative (OSCI) is currently defining a set of terminology for TLM and will eventually develop TLM standards. In reality, when engineers talk of TLM, they are probably talking about one or more of four different modeling styles that are discussed in the following section.

The underlying concept of TLM is to model only the level of detail that is needed by the engineers developing the system components and sub-system for a particular task in the development process. By modeling only the necessary details, design teams can realize huge gains in modeling speed thus enabling a new methodology. At this level, changes are relatively easy because the development team has not yet painted itself into a corner with low-level details such as a parallel bus implementation versus a serial bus implementation.

Using TLMs makes tasks usually reserved for hardware implementations practical to run on a model early in the system development process. TLM is a concept independent of language. However, to implement and refine TLM models, it is helpful to have a language like SystemC whose features support independent refinement of functionality and communication that is crucial to efficient TLM development.

Before exploring this new design methodology we will explore some of the background and terminology around TLM.

## 2.2 Abstraction Models

Several sets of terminology have been defined for the abstraction levels traditionally used in system models. We are presenting a slight variation of a model developed and presented by Dan Gajski and Lukai Cai at CODES (HW/SW Co-Design Conference) 2003 that is illustrated in *Figure 2-1*.

The first concept necessary for understanding TLM is that system and sub-system communication and functionality can be developed and refined independently. In this terminology, the communication and functionality components can be un-timed (UT), approximately-timed (AT), or cycle-timed (CT).

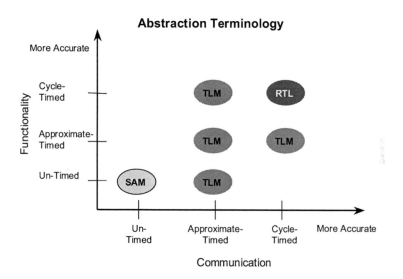

*Figure 2-1*. Abstraction Terminology

A model that is cycle-timing accurate for communication and for functionality is usually referred to as a register-transfer level (RTL) model. We refer to models with un-timed communication and functionality as a SAM. The RTL model is traditionally used for automatic synthesis to gates. Many times the SAM is used for algorithmic refinement and can be refined to approximately-timed communication and/or functionality.

The other four points plotted on the graph are usually collectively referred to as TLMs, and rely on approximately-timed functionality or communication. Approximately-timed models can rely on statistical timing, estimated timing, or sub-system timing requirements (or budgets) derived from system requirements.

A model with cycle-timed communication and approximately-timed functionality has been referred to as a Bus Functional Model(BFM) in older methodologies and the label is retained here. The three remaining TLMs have not yet developed commonly accepted names. For now, we will use the names developed by Gajski and Cai.

*Table 2-1.* Timing of Transaction-Level Models

| Model | Communication | Functionality |
|-------|---------------|---------------|
| SAM | UT | UT |
| Component assembly | UT | AT |
| Bus arbitration | AT | AT |
| Bus functional | CT | AT |
| Cycle-accurate computation | AT | CT |
| RTL | CT | CT |

All of these models are not necessary for most systems. In reality, most systems only need to progress through two or three points on the graph in Figure 2-1. With a language that supports refinement concepts, the transformation can be quite efficient.

## 2.3   Another Look at Abstraction Models

In this section, to build out your understanding of how TLM can be useful, we present a less rigorous and more example-based discussion of TLM. We will assume a generic system containing a microprocessor, a few devices, and memory connected by a bus.

The timing diagram in *Figure 2-2* illustrates one possible design outcome of a bus implementation. When first defining and modeling the system application, the exact bus-timing details do not affect the design decisions, and all the important information contained within the illustration is transferred between the bus devices as one event or transaction (component-assembly model).

Further into the development cycle, the number of bus cycles may become important (to define bus cycle-time requirements, etc.) and the information for each clock cycle of the bus is transferred as one transaction or event (bus-arbitration or cycle-accurate computation models).

When the bus specification is fully chosen and defined, the bus is modeled with a transaction or event per signal transition (bus functional or RTL model). Of course, as more details are added, more events occur and the speed of the model execution decreases.

In this diagram, the component assembly model takes 1 "event," the bus arbitration model takes approximately 5 "events," and the RTL model takes roughly 75 "events" (the exact number depends on the number of transitioning signals and the exact simulator algorithm). This simple example illustrates the magnitude of computation required and why more system design teams are employing a TLM-based methodology.

**Generic Bus Timing**

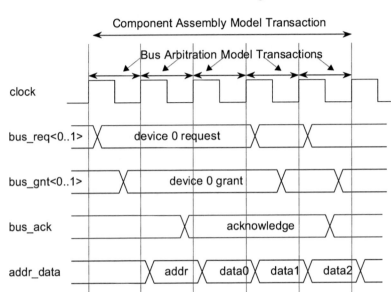

*Figure 2-2.* Generic Bus Timing Diagram

## 2.4   TLM-Based Methodology

Now that we have discussed some of the TLM concepts, we can look more closely at a TLM-based methodology as illustrated in *Figure 2-3*.

In this methodology, we still start with the traditional methods used to capture the customer requirements, a paper Product Requirements Document (PRD). Sometimes, the product requirements are obtained directly from a customer, but more likely the requirements are captured through the research of a marketing group.

From the PRD, a SAM is developed. The SAM development effort may cause changes or refinement to the PRD. The SAM is usually written by an architect or architecture group and captures the product specification or system critical parameters. In an algorithmic intensive system, the SAM will be used to refine the system algorithms.

The SAM is then refined into a TLM that may start as a component assembly type of TLM and is further refined to a bus arbitration model. The TLM is refined further as software design and development and hardware verification environment development progresses.

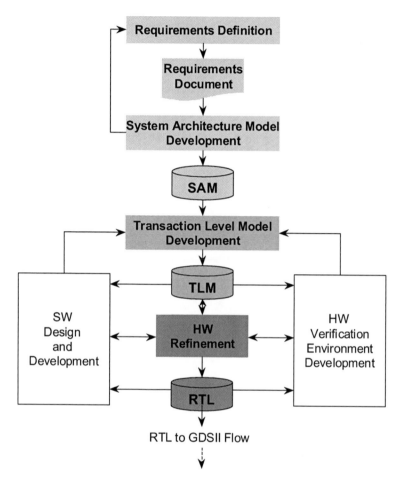

*Figure 2-3.* TLM-Based Flow

If the proper design language and techniques are used consistently throughout the flow, then the SAM can be reused and refined to develop the TLM. The TLM has several goals:

1. Refinement of implementation features such as HW/SW partitioning; HW partitioning among ASICs, FPGAs, and boards; bus architecture exploration; co-processor definition or selection; and many more
2. Development platform for system software
3. "Golden Model" for the hardware functional verification
4. Hardware micro-architecture exploration and a basis for developing detailed hardware specifications

In the near future, if EDA tools mature sufficiently, the TLM code may be refined to a behavioral synthesis model and be automatically converted to hardware from a higher-level abstraction than the current RTL synthesis flows. Today, the hardware refinement is likely done through a traditional paper specification and RTL development techniques, although the functional verification can now be performed via the TLM as outlined later in this chapter.

At first, development of the TLM appears to be an unnecessary task. However, the TLM creates benefits including:

- Earlier software development

- Earlier and better hardware functional verification test bench

- Creates a clear and unbroken path from customer requirements to detailed hardware and software specifications

After reading this book, you and your team should have the knowledge to implement TLMs quickly and effectively. The following section discusses in detail the benefits your team will bring to your organization when applying this methodology: early software development and early hardware functional verification.

### 2.4.1 Early Software Development

In complex systems where new software and new hardware are being created, software developers must often wait for the hardware design to be finalized before they can begin detailed coding. Software developers must also wait for devices (ICs and printed circuit boards) to be manufactured to test their code in a realistic environment. Even then, creating a realistic environment on a lab workbench can be very complex. This dependency creates a long critical path that may add so much financial risk to a project that it is never started.

*Figure 2-4* illustrates a traditional system development project schedule. The arrows highlight differences a TLM-based methodology would make. The time scale and the duration of each phase depend on the project size, project complexity, and the makeup of the system components (hardware, software, and algorithms).

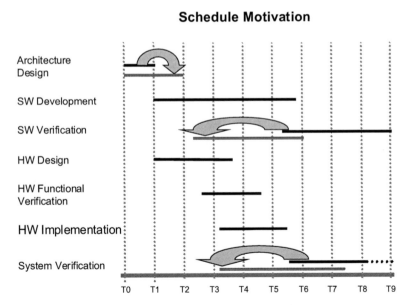

*Figure 2-4.* Schedule Benefits of Earlier Software Development

Creating a TLM from the SAM slightly lengthens the architectural design phase of a project, but it offers several potential benefits:

- Ability to start refining and testing software earlier, thereby reducing the overall development cycle

- Ability to provide earlier and more realistic hardware/software trade off studies at a time when changes are easier, thus improving overall system quality

- Ability to deliver executable models to customers both for validating the specification and driving changes, and acceleration of product adoption

- Ability to cancel (or redefine) an unrealistic project before spending even larger sums of money

Any opportunity to begin the software development work earlier warrants consideration. Indeed, the bottom line financial returns for just starting software development earlier, may dictate the adoption of this new methodology without the other benefits listed above.

### 2.4.2 Better Hardware Functional Verification

System design teams are always looking for ways to provide more and better functional verification of the hardware. The number of cases required to functionally verify a system is growing even faster than the actual system complexity.

Verifying the hardware interaction with the actual software and firmware before creating the hardware is becoming increasingly more important. With the chip mask set costs exceeding several hundred thousand dollars, finding out after making chips that a software workaround for the hardware is impossible or too slow is not acceptable. As a result, many teams are developing simulation and emulation techniques to verify the hardware interaction with the software and firmware.

Additionally, with the increase in size and complexity of the hardware, it is increasingly important to verify that unforeseen interactions within the chip, between chips, or between chips and software do not create unacceptable consequences. Debugging these interactions without significant visibility into the state of the chip being verified is very tough.

Very large Verilog or VHDL simulations along with emulation strategies have traditionally been used for system-level functional verification. With increasing system complexity, Verilog and VHDL simulations have become too slow for such verification. Hardware emulation techniques have been used when simulation has been too slow, but emulation techniques often have limited state visibility for debugging, and they can be very expensive.

When a design team develops a TLM, it is straightforward to refine the model to a verification environment through the use of adapters as outlined in the following section.

### 2.4.3 Adapters and Functional Verification

This section is a very brief overview of how a TLM model can be used as part of an overall system functional verification strategy. With modern systems, the hardware design is not fully debugged until it is successfully running the system software. This approach enables functional verification of the hardware with the system software prior to hardware availability. More details about implementation of this approach are given in Chapter 13, Custom Channels, and other sources[4].

To show one way that adapters can be applied to a TLM to create a verification environment, we will assume a generic system that looks like *Figure 2-5*. The generic system is composed of a microprocessor, memory, a couple of devices, and a bus with an arbiter.

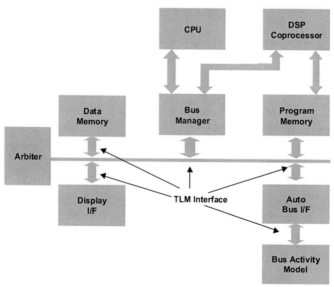

*Figure 2-5.* Generic System

---

[4] Grotker, T., Liao, S., Martin, G., Swan, S. 2002. *System Design with SystemC*. Norwell Massachusetts: Kluwer Academic Publishers.

For our discussions, we will concentrate on communication refinement and assume that the functionality of the devices, the memory, and the microprocessor will be approximately-timed or cycle-timed as appropriate throughout the design cycle.

In this very simple example, we assume that RTL views of the microprocessor and memory are not available or not important at this point in the verification strategy. In this case, the RTL for the two devices could be functionally verified by insertion of an adapter as illustrated in *Figure 2-6*.

This approach dictates that the adapter converts the timing-accurate signals of the bus coming from the RTL to a transaction view of the bus. The RTL sees the bus activity that would be created by the microprocessor, memory, and arbiter. Bus activity is propagated only to the non-RTL portion of the system after the adapter creates the transaction. This propagation creates a very high performance model compared to a traditional full RTL model.

This approach is just one way of applying adapters. The system-critical parameters, the system size, the system complexity, and more will contribute to a verification plan that will define a system-specific approach for application of adapters.

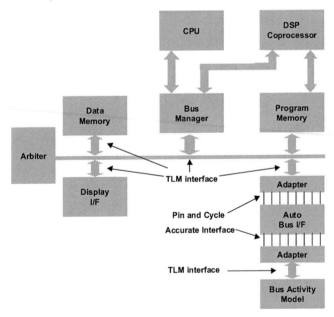

*Figure 2-6.* Adapter Example

## 2.5  Summary

A new TLM-based methodology is emerging to attack the design productivity of complex systems. The benefits of adopting this style of methodology are derived from early software development, early functional verification, and higher system quality. The productivity improvements derived from TLM-based methodology are huge and are the major motivation for adoption. Now, it is time to explore SystemC, a language that enables this new methodology.

# Chapter 3

# OVERVIEW OF SYSTEMC

Chapters 1 and 2 gave a brief context for the application of SystemC. This chapter discusses the SystemC language, but still presents an overview of the language elements. Details are discussed in-depth in subsequent chapters. Despite our best efforts not to use any part of the language before it is fully explained, some chapters may occasionally violate this goal due to the interrelated nature of SystemC. This chapter briefly discusses the major components of SystemC and their general usage and interactions as a way of giving context for the subsequent chapters. This chapter also provides this information as a brief overview for those times when unexplained SystemC components are required because of SystemC construct interactions.

The following diagram illustrates the major components of SystemC. To give a context throughout the book, we have included a duplicate of this diagram at the beginning of each new chapter. Bolded type indicates the topics discussed within that chapter.

| User libraries | SCV | Other IP |
|---|---|---|
| **Predefined Primitive Channels: Mutexs, FIFOs, & Signals** | | |

| SystemC | **Simulation Kernel** | **Threads & Methods** | **Channels & Interfaces** | **Data types: Logic, Integers, Fixed point** |
|---|---|---|---|---|
| | | **Events, Sensitivity & Notifications** | **Modules & Hierarchy** | |

| C++ | STL |
|---|---|

*Figure 3-1.* SystemC Language Architecture

For the rest of this chapter, we will discuss all of the components within the figure that are outlined in bold, but only after discussing the SystemC development environment and a few of the hardware-oriented features provided by SystemC. Much greater detail will be presented in subsequent chapters.

SystemC addresses the modeling of both software and hardware using C++. Since C++ already addresses most software concerns, it should come as no surprise that SystemC focuses primarily on non-software issues. The primary area of application for SystemC is the design of electronic systems, but SystemC has been applied to non-electronic systems[5].

## 3.1 C++ Mechanics for SystemC

We would like to start with the obligatory `Hello_SystemC` program but first let's look at the mechanics of compiling and executing a SystemC program or model.

As stated many times in this book, SystemC is a C++ class library, therefore, to compile and run a `Hello_SystemC` program, one must have a working C++ and SystemC environment.

The components of this environment include a:

- SystemC-supported platform
- SystemC-supported C++ compiler
- SystemC library (downloaded and compiled)
- Compiler command sequence, make file, or equivalent

The latest OSCI reference SystemC release (2.0.1 at this writing)[6] is available for free from www.systemc.org. The download contains scripts and make files for installation of the SystemC library as well as source code, examples, and documentation. The install scripts are compatible with the supported operating systems, and the scripts are relatively straightforward to execute by following the documentation.

The latest OS requirements can be obtained from the download in a readme file currently called `INSTALL`. SystemC is supported on various flavors of Sun Solaris, Linux, and HP/UX. At this time, the OS list is limited by the use of some assembly code that is used for increased simulation performance in the SystemC simulation kernel. The current release is also supported for various C++ compilers including GNU C++, Sun C++, and HP

---

[5] For example, read the book, *Microelectrofluidic Systems: Modeling and Simulation* by Tianhao Zhang et al., CRC Press, ISBN: 0849312760.

[6] SystemC version 2.1 should be available during the summer of 2004, and supports more platforms and compilers including MacOS X. Be sure to read the release notes carefully.

C++. The currently supported compilers and compiler versions can also be obtained from the INSTALL readme file in the SystemC download.

For beginners, this list should be considered exhaustive, although some hardy souls have ported various SystemC versions to other unsupported operating systems and C++ compilers. In addition, you will need gmake installed on your system to quickly compile and install the SystemC library with the directions documented in the INSTALL file.

The flow for compiling a SystemC program or design is very traditional, and is illustrated in *Figure 3-2* for GNU C++. Most other compilers will be similar. The C++ compiler reads each of the SystemC code file sets separately and creates an object file (usual file extension of .o). Each file set usually consists of two files typically with standard file extensions. We use .h and .cpp as file extensions, since these are the most commonly used in C++. The .h file is generally referred to as the header file and the .cpp file is often called the implementation file.

**Compilation Flow**

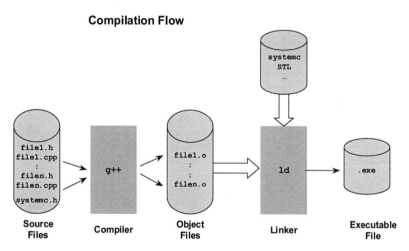

*Figure 3-2.* SystemC Compilation Flow

After creating the object files, the compiler (actually the loader or linker) will link your object files and the appropriate object files from the SystemC library (and other libraries such as the standard template library or STL). The resulting file is usually referred to as an executable, and it contains the SystemC simulation kernel and your design functionality.

The compiler and linker need to know two special pieces of information. First, the compiler needs to know where the SystemC header files are located (to support #**include** <systemc.h>). Second, the linker needs to know where the compiled SystemC libraries are located. This is typically accomplished by providing an environment variable named SYSTEMC, and ensuring the Makefile rules use the information.[7] If using gcc, the command probably looks something like this:

```
g++ -I$(SYSTEMC)/include \
    -L$(SYSTEMC)/lib-$(ARCH) -lsystemc \
    $(SRC)
```

*Figure 3-3.* Partial gcc Options to Compile and Link SystemC

The downloadable examples available from our website include Makefiles setup for Linux and gcc. Please refer to your C++ tool manuals for more information.

For the hardcore engineer types, you now have everything you need to compile and run a Hello_SystemC program; we have provided the obligatory program in *Figure 3-4*. Keywords for both C++ and SystemC are in bold. The rest of you now have an overview of how to compile and run the code examples in this book as well as your own SystemC creations. Everyone is now ready to dive into the language itself.

[7] For some installations, dynamic libraries may also be referenced if using the SystemC Verification library.

```cpp
#include <systemc.h>
#include <iostream>
SC_MODULE(Hello_SystemC) {//declare  the module class
  sc_in_clk iclk;              //define the clock port
  SC_CTOR(Hello_SystemC) {//create a constructor
    SC_METHOD(main_method);// register the main
                           // process
      sensitive << iclk.neg();//specify clk
                             // sensitivity
      dont_initialize();       //skip initial call
  }
  void main_method(void) {
    std::cout << sc_time_stamp()
              <<" Hello world!"
              << std::endl;
  }
};
int sc_main(int argc, char* argv[]) {
  //declare a time constant
  const sc_time t_PERIOD(8,SC_NS);
  //create periodic clock
  sc_clock clk("clk",t_PERIOD);
  //create an instance
  HelloWorld iHelloWorld("iHelloWorld");
  //connect the clock port and clock
  iHelloWorld.iclk(clk);
  // invoke the simulator
  sc_start(10,SC_NS);
  return 0;
}
```

*Figure 3-4.* Hello_SystemC Program Example

## 3.2   SystemC: A C++ Class for Hardware

SystemC provides mechanisms crucial to modeling hardware while using a language environment compatible with software development. SystemC provides several hardware-oriented constructs that are not normally available in a software language but are required to model hardware. All of the constructs are implemented within the context of the C++ language. This section looks at SystemC from the viewpoint of the hardware-oriented features. The major hardware-oriented features implemented within SystemC include:

- Time model

- Hardware data types

- Module hierarchy to manage structure and connectivity

- Communications management between concurrent units of execution

- Concurrency model

The following sections briefly discuss the implementation of these concepts within SystemC.

### 3.2.1 Time Model

SystemC tracks time with 64 bits of resolution using a class known as **sc_time**. Global time is advanced within the kernel. SystemC provides mechanisms to obtain the current time and implement specific time delays. To support ease of use, an enumerated type defines several natural time units from seconds (**SC_SEC**) to femtoseconds (**SC_FSEC**).

For those models that require a clock, a class called **sc_clock** is provided. This clock class is discussed in Chapter 14, Advanced Topics. The clock discussion is deferred to later chapters of the book, since many applications in SystemC do not require a clock (but do require a notion of time). Additionally, clocks do not add to the fundamental understanding of the language. By the later chapters, you should be able to implement the clock class yourself with the fundamentals learned throughout the book. However, you may find that you will still use the **sc_clock** class as a convenience.

### 3.2.2 Hardware Data Types

The wide variety of data types required by digital hardware are not provided inside the natural boundaries of C++ native data types, which are typically 8-, 16-, 32-, and 64-bit entities.

SystemC provides hardware-compatible data types that support explicit bit widths for both integral (e.g., `sc_int<>`) and fixed-point (e.g., `sc_fixed<>`) quantities. These data types are implemented using templated classes and generous operator overloading, so that they can be manipulated and used almost as easily as native C++ data types.

Furthermore, digital hardware can represent non-binary quantities such as tri-state and unknown state. These are supported with four-state logic (0,1,X,Z) data types (e.g., `sc_logic`). SystemC provides all the necessary methods for using hardware data types, including conversion between the hardware data types and conversion from hardware to software data types.

Finally, hardware is not always digital. SystemC does not currently directly support analog hardware; however, a working group has been formed to investigate the issues associated with modeling analog hardware in SystemC. For those with immediate analog issues, it is reasonable to model analog values using floating-point representations and providing the appropriate behavior.

### 3.2.3 Hierarchy and Structure

Large designs are almost always broken down hierarchically to manage complexity, easing understanding of the design for the engineering team. SystemC provides several constructs for implementing hardware hierarchy. Hardware designs traditionally use blocks interconnected with wires or signals for this purpose. For modeling hardware hierarchy, SystemC uses the module entity interconnected to other modules using channels. The hierarchy comes from the instantiation of module classes within other modules and is discussed in Chapter 10, Structure.

### 3.2.4 Communications Management

The SystemC channel provides a powerful mechanism for modeling communications, and the channel is one of the major contributions of SystemC version 2.0. Conceptually, a channel is more than a simple signal or wire. Channels can represent complex communications schemes that eventually map to significant hardware such as the AMBA bus[8]. At the same time, channels may also represent very simple communications such as a wire or a FIFO (first-in first-out queue). Channels are discussed in Chapters 8 Basic Channels, 9 Signals, and 13 Custom Channels.

The ability to have several quite different channel implementations used interchangeably to connect modules is a very powerful feature that enables an implementation of a "simple bus" replaced with a more detailed hardware implementation, and eventually implemented with gates. We briefly explore some of these concepts in Chapter 13, Custom Channels and in Chapter 14, Advanced Topics.

SystemC provides several built-in channels common to software and hardware design. These include **sc_mutex**, **sc_fifo**, **sc_signal<>** and others discussed later.

Finally, modules connect to channels via the port class, **sc_port<>**, a templated class that uses interface classes. Built-in interface classes include **sc_mutex_if**, **sc_fifo_in_if<>**, and others that are discussed fully in Chapters 10 Structure, 11 Connectivity, and 12 More on Ports.

### 3.2.5 Concurrency

Concurrency in a simulator is always an illusion. Simulators execute the code on a single physical processor. Even if you did have multiple processors performing the simulation, the number of units of concurrency in real hardware design will always outnumber the processors used to do the simulation by several orders of magnitude. Consider the problem of simulating the processors on which the simulator runs.

Simulation of concurrent execution is accomplished by simulating each concurrent unit (defined by an **SC_METHOD**, **SC_THREAD**, or **SC_CTHREAD**). Each unit is allowed to execute until simulation of the other units is required to keep behaviors aligned in time. In fact, the simulation code itself determines when the simulator makes these switches by the use of events. This simulation of concurrency is the same for SystemC, Verilog, VHDL, or

---

[8] See AMBA AHB Cycle Level Interface Specification at www.arm.com.

any other hardware description languages (HDL). In other words, the simulator uses a cooperative multi-tasking model. The simulator merely provides a kernel to orchestrate the swapping of the various concurrent elements, called processes. SystemC provides a simulation kernel that will be discussed lightly in the last section of this chapter. This kernel will be investigated more thoroughly in Chapter 6, A Notion of Time and in Chapter 9, Evaluate-Update Channels.

### 3.2.6 Summary of SystemC Features for Hardware Modeling

SystemC implements the structures necessary for hardware modeling by providing constructs that enable concepts of time, hardware data types, hierarchy and structure, communications, and concurrency. This section has presented an overview of SystemC relative to hardware design requirements for any design language. The following section discusses a complete set of language construct categories implemented in SystemC.

## 3.3 Overview of SystemC Components

In this section, we briefly discuss all the components of SystemC that are highlighted in *Figure 3-1*, which is the illustration that we will see at the beginning of each chapter throughout the book.

### 3.3.1 Modules and Hierarchy

Before getting started, it is necessary to have a firm understanding of two basic types of processes in SystemC. As indicated earlier, the SystemC simulation kernel schedules the execution of simulation processes. Simulation processes are simply member functions of **SC_MODULE** classes that are also "registered" with the simulation kernel.

Because the simulator kernel is the only caller of these member functions, they need no arguments, and they return no value. They are simply C++ functions that are declared as returning a void and having an empty argument list.

An **SC_MODULE** class can also have processes that are not executed by the simulation kernel, but are invoked as function calls within the simulation processes of the **SC_MODULE** class as normally done in C++.

### 3.3.2 Threads and Methods

From a software perspective, processes are simply threads of execution. From a hardware perspective, processes provide necessary modeling of independently-timed circuits. Simulation processes are member functions of an **SC_MODULE** that are registered with the simulation kernel. Registration occurs during the elaboration phase (during the execution of the constructor for the **SC_MODULE** class) using an **SC_METHOD**, **SC_THREAD**, or **SC_CTHREAD**[9] SystemC macro.

The most basic type of simulation process is known as the **SC_METHOD** process. This process is to be distinguished from the object-oriented concept of a class method or member function in C++. An **SC_METHOD** is simply a member function of an **SC_MODULE** class where time does not pass between the invocation and return of the function. In other words, an **SC_METHOD** is a purely normal function that happens to have no arguments and returns no value. A characteristic of a method process is that the simulator kernel repeatedly calls it.

The other basic type of simulation process is known as the **SC_THREAD**. This differs from the **SC_METHOD** in two ways. First, whereas **SC_METHOD**s are invoked multiple times, the **SC_THREAD** is only invoked *once*. Second, an **SC_THREAD** has the option to *suspend* itself and potentially allow time to pass before continuing. In this sense, an **SC_THREAD** is similar to a traditional software thread of execution.

There is a special case of the **SC_THREAD** known as the **SC_CTHREAD**. This process is simply a thread process that has the requirement of being sensitive to a clock. Sensitivity will be discussed later.

The **SC_METHOD**, **SC_THREAD**, and **SC_CTHREAD** are the basic units of concurrent execution. The simulation kernel invokes each of these processes, therefore they are generally never invoked directly by the user. The user indirectly controls execution of the simulation processes by the kernel as a result of events, sensitivity, and notification.

---

[9] **SC_CTHREAD** is under consideration for deprecation; however, several synthesis tools depend on it at the time of writing.

### 3.3.3 Events, Sensitivity, and Notification

Events, sensitivity, and notification are very important concepts for understanding the implementation of concurrency by the SystemC simulator.

Events are implemented by the SystemC **sc_event** class. Events are caused or fired through the **sc_event** member function, **notify**. The **notify** member function can occur within a simulation process (**SC_METHOD**, **SC_THREAD**, or **SC_CTHREAD**) or as a result of activity in a channel. When an **SC_METHOD**, **SC_THREAD**, or **SC_CTHREAD** process is sensitive to an event, and the event occurs, the simulation kernel schedules the process to be invoked.

SystemC has two types of sensitivity: static and dynamic. Static sensitivity is implemented by applying the SystemC **sensitive** command to an **SC_METHOD**, **SC_THREAD**, or **SC_CTHREAD** at elaboration time (within the constructor). Dynamic sensitivity lets a simulation process change its sensitivity on the fly. The **SC_METHOD** implements dynamic sensitivity with a **next_trigger**(*arg*) command. The **SC_THREAD** implements dynamic sensitivity with a **next**(*arg*) command. Both **SC_METHOD** and **SC_THREAD** can switch between dynamic and static sensitivity during simulation.

### 3.3.4 SystemC Data Types

Several hardware data types are provided in SystemC. Since the SystemC language is built on C++, all of the C++ data types are available, and the ability exists to define new data types for new hardware technology (i.e., multi-valued logic) or for applications other than electronic system design.

Hardware data types for mathematical calculations like **sc_fixed**<> and **sc_int**<> allow modeling of complex calculations like DSP functions and evaluate the performance when implemented in custom hardware or in processors without full floating-point capability.

Familiar data types like **sc_logic** and **sc_lv**<> are provided for RTL designers who need a data type to represent basic logic values or vectors of logic values.

### 3.3.5 Channels and Interfaces

Hardware designs typically contain hierarchy to reduce complexity. A block represents each level of hierarchy. VHDL refers to blocks as entity/architecture pairs, which separate an interface specification from the body of code for each block. In Verilog, blocks are called modules, and contain both interface and implementation in the same code. SystemC separates the interface/implementation similar to VHDL. The C++ notion of header (.h file) is used for the entity and the notion of implementation (.cpp file) is used for the architecture.

Blocks communicate via ports/pins and signals or wires in traditional HDLs. In SystemC, modules are interconnected using either primitive channels or hierarchical channels. Both types of channels connect to modules via ports. The powerful ability to have interchangeable channels is implemented through a component called an interface.

Interestingly, module interconnection happens programmatically in SystemC, during the elaboration phase. This interconnection lets designers build regular structures using loops and conditional statements (see Chapter 14, Advanced Topics). From a software perspective, elaboration is simply the period of time when modules invoke their constructor methods. Currently, SystemC only allows construction prior to the start of simulation.

### 3.3.6 Summary of SystemC Components

Now, it is time to tie all of these basic concepts together into one illustration, *Figure 3-5*. This illustration is used many times throughout the book when referring to the different SystemC components. It can appear rather intimidating since it shows almost all of the concepts within one diagram. In practice, most **SC_MODULE**s will not contain all of the illustrated components.

## SystemC Components

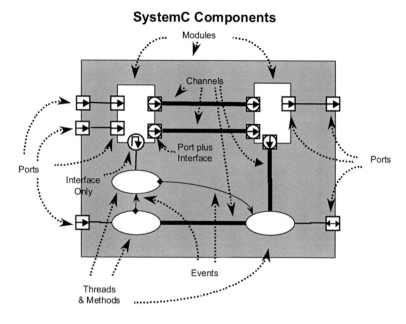

*Figure 3-5.* SystemC Components

The figure shows the concept of an **SC_MODULE** that can contain instances of other **SC_MODULE**s. An **SC_METHOD**, **SC_THREAD**, or **SC_CTHREAD** can also be defined within an **SC_MODULE**.

Communication between modules and **SC_METHOD**s, **SC_THREAD**s, and **SC_CTHREAD**s is accomplished through various combinations of ports, interfaces, and channels. Coordination between simulation processes (**SC_METHOD**, **SC_THREAD**, **SC_CTHREAD**) is accomplished through events.

The rest of this book describes all of these components and their interaction with the SystemC simulation kernel in detail.

## 3.4   SystemC Simulation Kernel

The SystemC simulator has two major phases of operation: *elaboration* and *execution*. A third, often minor, phase occurs at the end of execution, and could be characterized as post-processing or *cleanup*.

Execution of statements prior to the **sc_start()** function call is known as the elaboration phase. This phase is characterized by the initialization of data structures, the establishment of connectivity, and the preparation for the second phase, execution.

The execution phase hands control to the SystemC simulation kernel, which orchestrates the execution of processes to create an illusion of concurrency.

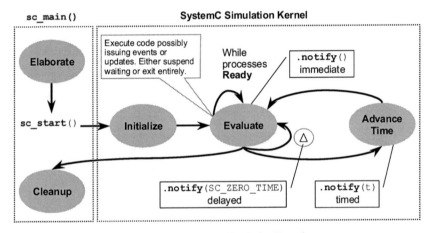

*Figure 3-6.* SystemC Simulation Kernel

The illustration in *Figure 3-6* should look very familiar to those who have studied Verilog and VHDL simulation kernels. Very briefly, after **sc_start()** all simulation processes (minus a few exceptions) are randomly invoked during initialization. After initialization, a simulation process is run when an event to which it is sensitive occurs. Several simulation processes may begin at the same instant in simulator time. Because of this case, all of the simulation processes are evaluated and then their outputs are updated. An evaluation followed by an update is referred to as a *delta-cycle*. If no additional simulation processes need to be evaluated at that instant (as a result of the update), then simulation time is advanced. When no additional simulation processes need to run, the simulation ends.

This brief overview of the simulation kernel is meant to give you an overview for the rest of the book. This diagram will be used again to explain important intricacies later. It is very important to understand how the kernel functions to fully understand the SystemC language. We have provided an animated version of this diagram walking through a small code example at our website, www.EklecticAlly.com. The SystemC LRM (Language Reference Manual) specifies the behavior of the SystemC simulation kernel, and is the definitive source. We encourage the reader to use any or all of these resources during their study of SystemC to fully understand the simulation kernel.

Chapter 4

# DATA TYPES

SystemC has a number of predefined data types to support hardware designs spanning from the native C++ data types to specialized fractional fixed-point representations. Choosing a data type depends on the range of values to be represented, the required precision, and the required operations. Choice of a data type also affects the speed of simulation, synthesizability, and synthesis results. The data types used differ depending on the level of abstraction represented in your model.

## 4.1 Numeric Representation

Representation of literal data is fundamental to all languages. C++ allows for simple integers, floats, Booleans, characters, and strings.

To support hardware data representations, SystemC provides a unified string representation using C-style strings. It is possible to convert both to and from this format. SystemC uses the following syntax for strings:

```
sc_string name("0 base [sign ] number [e[+|-] exp]");
// no whitespace    Zero
```

*Figure 4-1.* Syntax of sc_string

Where *base* is one of **b**, **o**, **d**, or **x** for binary, octal, decimal, and hexadecimal, respectively. The *sign* allows specification of signed (empty), unsigned (**us**), signed magnitude (**sm**), and canonical signed digit (**csd**) numbers. *number* is an integer in the indicated base. The optional exponent *exp* is always specified using decimal. Eleven specific representations from the SystemC LRM are shown in the table below. Notice the enumeration column, **sc_numrep**, which is used when converting into a unified string.

*Table 4-1.* Unified String Representation for SystemC

| sc_numrep | Prefix | Meaning | sc_int<5>(-13) [10] |
|-----------|--------|---------|---------------------|
| SC_DEC | 0d | Decimal | "-0d13" |
| SC_BIN | 0b | Binary | "0b10011" |
| SC_BIN_US | 0bus | Binary unsigned | "0bus01101" |
| SC_BIN_SM | 0bsm | Binary signed magnitude | "-0bsm01101" |
| SC_OCT | 0o | Octal | "0o03" |
| SC_OCT_US | 0ous | Octal unsigned | "0ous15" |
| SC_OCT_SM | 0osm | Octal signed magnitude | "-0osm03" |
| SC_HEX | 0x | Hex | "0xf3" |
| SC_HEX_US | 0xus | Hex unsigned | "0xus0d" |
| SC_HEX_SM | 0xsm | Hex signed magnitude | "-0xsm0d" |
| SC_CSD | 0csd | Canonical signed digit | "0csd-010-" |

Here are some examples of literal data in SystemC:

```
sc_string a ("0d13"); // decimal 13
a = sc_string ("0b101110"); // binary of decimal 44
```

*Figure 4-2.* Example of sc_string

---

[10] +13 for unsigned types

## 4.2 Native Data Types

SystemC supports all the native C++ data types: `int`, `long int`, `int`, `unsigned int`, `unsigned long int`, `unsigned short int`, `short`, `double`, `float`, `char`, and `bool`.

For many SystemC designs, the built-in C++ data types should be sufficient. Native C++ data types are the most efficient in terms of memory usage and execution speed of the simulator.

```
// Example native C++ data types
int             spark_offset; // Adjustment for
                              // ignition
unsigned        repairs = 0;  // Count repair
                              // incidents
unsigned long   mileage;      // Miles driven
short int       speedometer;  // -20..0..100 MPH
float           temperature;  // Engine temp in C
double          time_of_last_request; //Time of bus
                                      //activity
std::string     license_plate;// Text for license
                              // plate
const bool      WARNING_LIGHT = true;// Status
                                     // indicator
// Direction of travel
enum            compass   {SW,W,NW,N,NE,E,SE,S};
```

*Figure 4-3.* Example of C++ Built-In Data Types

## 4.3 Arithmetic Data Types

SystemC provides two sets of numeric data types. Arithmetic operations may be performed on numeric data. One set models data with bit widths up to 64-bits wide; another set models data with bit widths larger than 64-bits wide. Most of the native C++ data types have widths, ranges, and interpretations that are compiler-implementation-defined to match the host computer for execution efficiency.

### 4.3.1 sc_int and sc_uint

Most hardware needs to specify actual storage width at some level of refinement. When dealing with arithmetic, the built-in **sc_int** and **sc_uint** (unsigned) numeric data types provide an efficient way to model data with specific widths from 1- to 64-bits wide. When modeling numbers with data whose width is not an integral multiple of the simulating processor's data paths, some bit masking and shifting must be performed to fit internal computation results into the declared data format.

Thus, any data type that is not native to both the C++ language and the processor width will simulate slower than the native types. Thus, built-in C++ data types are faster than **sc_int** and **sc_uint**.

```
sc_int<LENGTH>   NAME…;
sc_uint<LENGTH>  NAME…;
```

*Figure 4-4.* Syntax of Arithmetic Data Types

Significant speed improvements can be attained if all **sc_int**s are 32 or fewer bits by simply setting the **-D_32BIT_** compiler flag.

GUIDELINE    Do not use **sc_int** unless or until prudent. One necessary condition for using **sc_int** is when using synthesis tools that require hardware representation.

### 4.3.2 sc_bigint and sc_biguint

Some hardware may be larger than the numbers supported by native C++ data types. SystemC provides **sc_bigint** and **sc_biguint** for this purpose. These data types provide large number support at the cost of speed.

```
sc_bigint<BITWIDTH>   NAME…;
sc_biguint<BITWIDTH>  NAME…;
```

*Figure 4-5.* Syntax of sc_bigint and sc_biguint

```
// SystemC integer data types
sc_int<5>      seat_position=3; //5 bits: 4 plus sign
sc_uint<13>    days_SLOC(4000); //13 bits: no sign
sc_biguint<80> revs_SLOC;       // 80 bits: no sign
```

*Figure 4-6.* Example of SystemC Integer Data Types

GUIDELINE: Do not use **sc_bigint** for 64 or fewer bits. Doing so causes performance to suffer compared to using **sc_int**.

## 4.4 Boolean and Multi-Value Data Types

SystemC provides one set of data types for Boolean values and another set of data types for unknown and tri-state values.

### 4.4.1 sc_bit and sc_bv

For ones and zeroes, SystemC provides the **sc_bit**, and for long bit vectors SystemC provides **sc_bv**<> (bit vector) data types. These types do not support arithmetic data like the **sc_int** types, and these data types don't execute as fast as the built-in **bool** and Standard Template Library **bitset** types. As a result, these data types are being considered for deprecation (i.e., removal from the language).

```
sc_bit            NAME...;
sc_bv<BITWIDTH>  NAME...;
```

*Figure 4-7.* Syntax of Boolean Data Types

**sc_bit** and **sc_bv** come with some supporting data constants, **SC_LOGIC_1** and **SC_LOGIC_0**. For less typing, if **using namespace sc_dt**, then type **Log_1** and **Log_0**, or even type '1' and '0'.

Operations include the common bitwise **and, or, xor** operators (i.e., **&, |, ^**). In addition to bit selection and bit ranges (i.e., **[]** and **range**()), **sc_bv**<> also supports **and_reduce**(), **or_reduce**(), **nand_reduce**(), **nor_reduce**(), **xor_reduce**(), and **xnor_reduce**() operations. Reduction operations place the operator between all adjacent bits.

```
sc_bit flag(SC_LOGIC_1); // more efficient to use bool
sc_bv<5> positions = "01101";
sc_bv<6> mask = "100111";
sc_bv<5> active = positions & mask;// 00101
sc_bv<1> all = active.and_reduce (); // SC_LOGIC_0
positions.range (3,2) = "00";// 00001
positions[2] = active[0] ^ flag;
```

*Figure 4-8.* Examples of bit operations

### 4.4.2 sc_logic and sc_lv

More interesting than the Boolean data types are the multi-value data types used to represent unknown and high impedance (i.e., tri-state) conditions. SystemC represents these with the **sc_logic** and **sc_lv<>** (logic vector) data types. These types are represented with **SC_LOGIC_1**, **SC_LOGIC_0**, **SC_LOGIC_X**, and **SC_LOGIC_Z**. For less typing, if **using namespace sc_dt**, then type **Log_1**, **Log_0**, **Log_X**, and **Log_Z**, or even type '1', '0', 'X' and 'Z'.

Because of the overhead, these data types are considerably slower than their **sc_bit** and **sc_bv** counterparts. For best performance, always use built-in types such as **bool**.

```
  sc_logic        NAME [,NAME]...;
  sc_lv<BITWIDTH> NAME[,NAME]...;
```

*Figure 4-9.* Syntax of Multi-Value Data Types

SystemC does not represent other multi-level data types or drive strengths like Verilog's 12-level logic values or VHDL's 9-level **std_logic** values. However, you can create custom data types if truly necessary, and you can manipulate them by operator overloading in C++.

```
sc_logic buf(sc_dt::Log_Z);
sc_lv<8> data_drive("zz01XZ1Z");
data_drive.range (5,4) = "ZZ";// ZZZZXZ1Z
buf = '1';
```

*Figure 4-10.* Examples of 4-Level Logic Types

## 4.5 Fixed-Point Data Types

SystemC provides the following fixed-point data types: **sc_fixed**, **sc_ufixed**, **sc_fix**, **sc_ufix**, and the **_fast** variants of them.

Integral data types do not satisfy all design types. In particular, DSP applications often need to represent numbers with fractional components. SystemC provides eight data types providing fixed-point numeric representation. While native **float** and **double** data types satisfy high-level representations, realizable hardware has speed and area requirements. Also, integer-based DSP processors do not natively support floating point. Fixed-point numbers provide an efficient solution[11] in both hardware and software.

A number of parameters that control fixed-point behavior (e.g., overflow and underflow) must be set. What follows will briefly cover these aspects, but for full information, please see the SystemC LRM.

IMPORTANT: To improve compile times, the SystemC header code omits fixed-point data types unless the **#define SC_INCLUDE_FX** is specified prior to **#include** <systemc.h> in your code.

```
sc_fixed<WL,IWL[,QUANT[,OVFLW[,NBITS]]>        NAME...;
sc_ufixed<WL,IWL[,QUANT[,OVFLW[,NBITS]]>       NAME...;
sc_fixed_fast<WL,IWL[,QUANT[,OVFLW[,NBITS]]>   NAME...;
sc_ufixed_fast<WL,IWL[,QUANT[,OVFLW[,NBITS]]>  NAME...;

sc_fix_fast      NAME(WL,IWL[,QUANT[,OVFLW[,NBITS]])...;
sc_ufix_fast     NAME(WL,IWL[,QUANT[,OVFLW[,NBITS]])...;
sc_fixed_fast    NAME(WL,IWL[,QUANT[,OVFLW[,NBITS]])...;
sc_ufixed_fast   NAME(WL,IWL[,QUANT[,OVFLW[,NBITS]])...;
```

*Figure 4-11.* Syntax of Fixed-Point Data Types

These data types have several easy-to-remember distinctions. First, those ending with **_fast** are faster than the others are because their precision is

---

[11] At the time of writing, fixed-point data types were not synthesizable; however, private discussions indicate some EDA vendors are considering this as a possible new feature.

limited to 53 bits. Internally, **_fast** types are implemented using C++ **double**[12].

Second, the prefix **sc_ufix** indicates unsigned just as **uint** are unsigned integers. Third, the past tense **ed** suffix to **fix** indicates a templated data type that must have static parameters defined using compile-time constants.

Remember that **fixed** is past tense (i.e., already set in stone), and it cannot be changed after compilation. On the other hand, you can dynamically change those data types lacking the past tense (i.e., the **_fix** versions). Non-**ed** types are still active, and you can change them on the fly.

The parameters needed for fixed-point data types are the word length (*WL*), integer-word length (*IWL*), quantization mode (*QUANT*), overflow mode (*OVFLW*), and number of saturation bits (*NBITS*). Word length (*WL*) and integer word length (*IWL*) have no defaults and need to be set.

The word length establishes the total number of bits representing the data type. The integer word length indicates where to place the binary decimal point and can be positive or negative. *Figure 4-12* below illustrates how this works.

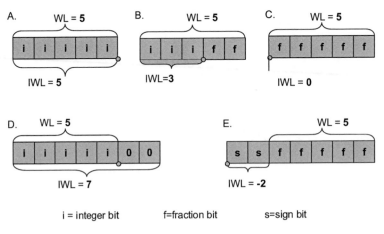

*Figure 4-12.* Fixed-Point Formats

---

[12] This implementation takes advantage of the linearly scaled 53-bit integer mantissa inside a 64-bit IEEE-754 compatible floating-point unit. On processors without an FPU, this behavior must be emulated in software, and there will be no speed advantage.

The *Figure 4-12*shows examples with the binary point in several positions. Consider example B in the preceding figure. This could be declared as **sc_fixed**<5, 3>, and would represent values from –4.75 up to 3.75 in 1/4 increments.

You can select several overflow modes from a set of enumerations that are listed in the next table. A similar table for the quantization modes is also shown. Overflow mode, quantization mode, and number of saturation bits all have defaults. You can establish defaults by setting up **sc_fxtype_context** objects for the non-**ed** data types.

*Table 4-2.* Overflow Mode Enumerated Constants

| Name | Overflow Meaning |
|------|------------------|
| **SC_SAT** | Saturate |
| **SC_SAT_ZERO** | Saturate to zero |
| **SC_SAT_SYM** | Saturate symmetrically |
| **SC_WRAP** | Wraparound |
| **SC_WRAP_SYM** | Wraparound symmetrically |

*Table 4-3.* Quantization Mode Enumerated Constants

| Name | Quantization Mode |
| --- | --- |
| **SC_RND** | Round |
| **SC_RND_ZERO** | Round towards zero |
| **SC_RND_MIN_INF** | Round towards minus infinity |
| **SC_RND_INF** | Round towards infinity |
| **SC_RND_CONV** | Convergent rounding[13] |
| **SC_TRN** | Truncate |
| **SC_TRN_ZERO** | Truncate towards zero |

The following examples should help explain the syntax for the fixed-point data types.

```
const sc_ufixed<19,3> PI("3.141592654");
sc_fix oil_temp(20,17,SC_RND_INF,SC_SAT);
sc_fixed_fast<7,1> valve_opening;
```

*Figure 4-13.* Examples of Fixed-Point Data Types

Only the word length and integer word length are required parameters. If not specified, the default overflow is **SC_WRAP**, the default quantization is **SC_TRN**, and saturation bits defaults to one.

A special note applies if you intend to set up arrays of the **_fix** types. Since a constructor is required, but C++ syntax does not allow arguments for this situation, it is necessary to use the **sc_fxtype_context** type to establish the defaults.

For significantly more information, refer to section 6.8 of the SystemC LRM.

---

[13] Convergent rounding is probably the oddest. If the most significant deleted bit is one, and either the least significant of the remaining bits or at least one of the other deleted bits is one, then add one to the remaining bits.

## 4.6 Operators for SystemC Data Types

The SystemC data types support all the common operations with operator overloading.

*Table 4-4.* Operators

| | |
|---|---|
| **Comparison** | `== != > >= < <=` |
| **Arithmetic** | `++ -- * / % + -` |
| **Bitwise** | `~ & | ^` |
| **Assignment** | `= &= |= ^= *= /= %= += -= <<= >>=` |

In addition, SystemC provides special methods to access bits, bit ranges, and perform explicit conversions.

*Table 4-5.* Special Methods

| | |
|---|---|
| **Bit Selection** | `bit(`*idx*`), [`*idx*`]` |
| **Range Selection** | `range(`*high,low*`), (`*high,low*`)` |
| **Conversion** (to C++ types) | `to_double(), to_int(),`<br>`to_int64(), to_long(),`<br>`to_uint(), to_uint64(),`<br>`to_ulong(), to_string(`*type*`)` |
| **Testing** | `is_zero(), is_neg(), length()` |
| **Bit Reduction** | `and_reduce(), nand_reduce(),`<br>`or_reduce(), nor_reduce(),`<br>`xor_reduce(), xnor_reduce()` |

In general, all the common combinations you would expect are present. For more information, refer to the SystemC LRM.

One often overlooked aspect of these data types (and C++ data types) is mixing types in arithmetic operations. It is OK to mix similar data types of different lengths, but crossing types is dangerous. For example, assigning the results of an operation involving two **sc_int**s to an **sc_bigint** does not automatically promote the operand to **sc_bigint** for intermediate calculations. To accomplish that, it is necessary to have one of the arguments be an **sc_bigint** or perform an explicit conversion of one of at least one of the operand arguments. Here is an example (addition):

```
sc_int<3> d(3);
sc_int<5> e(15);
sc_int<5> f(14);
sc_int<7> sum = d + e + f;// Works
sc_int<64> g("0x7000000000000000");
sc_int<64> h("0x7000000000000000");
sc_int<64> i("0x7000000000000000");
sc_bigint<70> bigsum = g + h + i; // Doesn't work
bigsum = sc_bigint<70>(g) + h + i;// Works
```

*Figure 4-14.* Example of Conversion Issues

## 4.7 Higher Levels of Abstraction and the STL

The basic C++ and SystemC data types lack structure and hierarchy. For these, the standard C++ **struct** and array are good starting points. However, a number of very useful data type classes are freely available, which provides another benefit of having a modeling language based upon C++.

The Standard Template Library (STL) is the most popular of these libraries, and it comes with all modern C++ compilers. The STL contains many useful data types and structures, including an improved character array known as **string**. This book will not attempt to cover the STL in any detail, but a brief overview may stimulate you to search further.

The STL has generic containers such as the **vector<>, map<>, list<>**, and **deque<>**, which may contain various data types. These containers can be manipulated by STL algorithms such as **for_each()**, **count()**, **min_element()**, **max_element()**, **search()**, **transform()**, **reverse()**, and **sort()**. These are just a few of the algorithms available.

The STL container **vector**<> closely resembles the common C++ array, but with several useful improvements. First, the **vector**<> may be resized dynamically. Second, and perhaps more importantly, accessing an element of a **vector**<> can have bounds checking for safety. The example below demonstrates use of an STL **vector**<>.

```
#include <vector>
int main(int argc, char* argv[]) {
  std::vector<int> mem(1024);
  for (unsigned i=0; i!= 1024; i++) {
    // Following checks access (safer than mem[I])
    mem.at(i) = -1; // initialize memory to known
                    // values
  }//endfor
  ...
  mem.resize(2048); // increase size of memory
  ...
}//end main()
```

*Figure 4-15.* Example of STL Vector

Large sparsely-used memories occupy too much space when implemented as arrays or vectors. These memories require the attributes of an associative map such as the STL **map**<>. A **map**<> requires specification of both the index and the value data types. Only index values that you have assigned occupy storage space. Thus, you can represent a large data space with minimal real memory.

```
#include <iostream>
#include <map>
int main(int argc, char* argv[]) {
  typedef unsigned long ulong;
  std::map<ulong, int> lmem; //possible 2^64
                              //locations!
  // Fill ten random locations with random values
  while (lmem.size() < 10) {
    // 10 random memory location/values
    lmem[rand()]=rand();
  }//endwhile
  // Display memory contents
  typedef std::map<ulong,int>::const_iterator iter;
  for (iter iv=lmem.begin();iv!=lmem.end();++iv) {
    std::cout << std::hex
      << "lmem[" << iv->first
      << "]=" << iv->second << ";" << std::endl;
  }//endfor
}//end main()
```

*Figure 4-16.* Example of an STL Map

## 4.8 Choosing the Right Data Type

A frequent question is, "Which data types should be used for this design?" The best answer is, "Choose a data type that is closest to native C++ as possible for the modeling needs at hand." Choosing native data types will always produce the fastest simulation speeds.

The table below gives an idea of performance.

*Table 4-6.* Data Type Performance

| | |
|---|---|
| Fastest | Native C/C++ Data Types (e.g., **int**, **double** and **bool**) |
| | **sc_int<>, sc_uint<>** |
| | **sc_bit, sc_bv<>** |
| | **sc_logic, sc_lv<>** |
| | **sc_bigint<>, sc_biguint<>** |
| | **sc_fixed_fast<>, sc_fix_fast, sc_ufixed_fast<>, sc_ufix_fast** |
| Slowest | **sc_fixed<>, sc_fix, sc_ufixed<>, sc_ufix** |

Some types of modeling tools may impose requirements on data types. For instance, RTL synthesis tools generally require all data to be in **sc_*** data types, and do not synthesize floating-point or fixed-point data types.

## 4.9 Exercises

For the following exercises, use the samples provided at www.EklecticAlly.com/Book/.

**Exercise 4.1:** Examine, compile, and run the examples from the website, datatypes and uni_string_rep. Note that although these examples include **systemc.h**, they only use data types.

**Exercise 4.2:** Write a program to read data from a file using the unified string representation and store in an array of **sc_uint**. Output the values as **SC_DEC** and **SC_HEX_SM**.

**Exercise 4.3:** Write a program to generate 100,000 random values and compute the squares of these values. Do the math using each of the following data types: **short, int, unsigned, long, sc_int<8>, sc_uint<19>, sc_bigint<8>, sc_bigint<100>, sc_fixed<12,12>**. Be certain to generate numbers distributed over the entire range of possibilities. Compare the run times of each data type.

**Exercise 4.4:** Examine, compile, and run the example addition. What would it take to fix the problems noted? Try adding various sizes of **sc_bigint<>**.

| Predefined Primitive Channels: Mutexes, FIFOs, & Signals | | | |
|---|---|---|---|
| Simulation Kernel | Threads & Methods | Channels & Interfaces | Data types: Logic, Integers, Fixed point |
| | Events, Sensitivity & Notifications | **Modules & Hierarchy** | |

# Chapter 5

# MODULES
*SC_MODULE*

This chapter lays the foundation for SystemC models. In this chapter, we explore how to put together a minimal SystemC program in preparation for an exploration of time and concurrency in the later chapters. With respect to hierarchy, this chapter only touches the very top level. Chapter 10, Structure will discuss hierarchy in more detail.

## 5.1 A Starting Point: sc_main

All programs need a starting point. In C/C++, the starting point is called **main**. In Verilog and VHDL, it might superficially appear that every process starts at once. In reality, some time passes between initializing the code and beginning the simulation. SystemC having its roots in C/C++ exposes the starting point, and that starting point is known as **sc_main**.

The top level of a C/C++ program is a function named **main()**. Its declaration is generally:

```
int main(int argc, char* argv[]) {
    BODY_OF_PROGRAM
  return EXIT_CODE; // Zero indicates success
}
```

*Figure 5-1.* Syntax of C++ main()

In the figure above, argc represents the number of command-line arguments including the program name itself. The second argument, argv[], is an array of C-style character strings representing the command line that invoked the program. Thus, argv[0] is the program name itself.

SystemC usurps this procedure and provides a replacement known as **sc_main()**. The SystemC library provides its own definition of **main()**, which in turn calls **sc_main**() and passes along the command-line arguments. Thus, the form of **sc_main**() follows:

```
int sc_main(int argc, char* argv[]) {
  ELABORATION
  sc_start(); // <-- Simulation begins & ends
              //     in this function!
  [POST-PROCESSING]
  return EXIT_CODE; // Zero indicates success
}
```

*Figure 5-2.* Syntax of sc_main()

By convention, SystemC programmers simply name the file containing **sc_main**(), as main.cpp to indicate to the C/C++ programmer that this is the place where everything begins[14]. The actual **main**() routine is located in the SystemC library itself.

Within **sc_main**(), code executes in three distinct major phases. Let us examine these phases, which are elaboration, simulation, and post-processing.

During elaboration, structures needed to describe the interconnections of the system are connected. Elaboration establishes hierarchy and initializes the data structures. Elaboration consists of creating instances of clocks, design modules, and channels that interconnect designs. Additionally, elaboration invokes code to register processes and perform the connections between design modules. Within each design, additional layers of design hierarchy are possible.

At the end of elaboration, **sc_start**() invokes the simulation phase. During simulation, code representing the behavior of the model executes. Chapter 7, Concurrency will explore this phase in detail.

Finally, after returning from **sc_start**(), the post-processing phase begins. Post-processing is mostly optional. During post-processing, code may read data created during simulation and format reports or otherwise handle the results of simulation.

Post-processing finishes with the **return** of an exit status from **sc_main**(). A non-zero return status indicates failure, which may be a computed result of post-processing. A zero return should indicate success (i.e., confirmation that the model correctly passed all tests). Many coders

---

[14] This naming convention is not without some controversy in some programming circles; however, most groups have accepted it and deal with the name mismatch

neglect this aspect and simply return zero by default. We recommend that you explicitly confirm the model passed all tests.

We now turn our attention to the components used to create a system model.

## 5.2  The Basic Unit of Design: SC_MODULE

Complex systems consist of many independently functioning components. These components may represent hardware, software, or any physical entity. Components may be large or small. Components often contain hierarchies of smaller components. The smallest components represent behaviors and state. In SystemC, we use a concept known as the **SC_MODULE** to represent components.

DEFINITION: A SystemC module is the smallest container of functionality with state, behavior, and structure for hierarchical connectivity.

A SystemC module is simply a C++ class definition. Normally, a macro **SC_MODULE** is used to declare the class:

```
#include <systemc.h>
SC_MODULE(module_name) {
  MODULE_BODY
};
```

*Figure 5-3.* Syntax of SC_MODULE

**SC_MODULE** is a simple cpp[15] macro:

```
#define SC_MODULE(module_name) \
        struct module_name: public sc_module
```

*Figure 5-4.* SystemC Header Snippet of SC_MODULE as #define

---

[15] cpp is the C/C++ pre-processor that handles # directives such as **#define**.

Within this derived module class, a variety of elements make up the
*MODULE BODY*:

- Ports
- Member channel instances
- Member data instances
- Member module instances (sub-designs)
- Constructor
- Destructor
- Process member functions (processes)
- Helper functions

Of these, only the constructor is required; however, to have any useful
behavior, you must have either a process or a sub-design. Let us first look at
the constructor in Section 5.3, and then a simple process in Section 5.4. This
sequence lets us finish in Section 5.5 with a basic example of a minimal
design.

## 5.3 The SC_MODULE Class Constructor: SC_CTOR

The **SC_MODULE** constructor performs several tasks specific to SystemC.
These tasks include:

- Initializing/allocating sub-designs (Chapter 10, Structure)
- Connecting sub-designs (Chapters 10, Structure and 11, Connectivity)
- Registering processes with the SystemC kernel (Chapter 7, Concurrency)
- Providing static sensitivity (Chapter 7, Concurrency)
- Miscellaneous user-defined setup

To simplify coding, SystemC provides a C-preprocessor (cpp) macro, **SC_CTOR**(). The syntax of this macro follows:

```
SC_CTOR(module_name)
: Initialization // OPTIONAL
{
  Subdesign_Allocation
  Subdesign_Connectivity
  Process_Registration
  Miscellaneous_Setup
}
```

*Figure 5-5.* Syntax of SC_CTOR

Let us now examine the process, and see how it fits.

## 5.4 The Basic Unit of Execution: SystemC Process

The process is the basic unit of execution in SystemC. From the time the simulator begins until simulation ends, all executing code is initiated from one or more processes. Processes appear to execute concurrently.

DEFINITION: A SystemC process is a member function or class method of an **SC_MODULE** that is invoked by the scheduler[16] in the SystemC simulation kernel.

The prototype of a process member function for SystemC is:

```
void PROCESS_NAME(void[17]);
```

*Figure 5-6.* Syntax of SystemC Process

A SystemC process takes no arguments and returns none. This syntax makes it simple for the simulation kernel to invoke. There are several kinds of processes, and we will discuss all of them eventually. For the purposes of simplification, we will look at only one process type in this chapter[18].

---

[16] We will look closely at the scheduler in Chapter 7, Concurrency.

[17] The keyword void is optional here, and it is typically left out.

[18] Chapter 7, Concurrency looks at the more common process types (e.g., **SC_THREAD** and **SC_METHOD**) in more detail, and Chapter 14 Advanced Topics finishes out the discussion of processes with less common types (e.g., **SC_CTHREAD** and dynamic processes).

The easiest type of process to understand is the SystemC thread, **SC_THREAD**. Conceptually, a SystemC thread is identical to a software thread. In simple C/C++ programs, there is only one thread running for the entire program. The SystemC kernel lets many threads execute in parallel, as we shall learn in Chapter 7, Concurrency.

A simple **SC_THREAD** begins execution when the scheduler calls it and ends when the thread exits or returns. An **SC_THREAD** is called only once[19], just like a simple C/C++ program. An **SC_THREAD** may also suspend itself, but we will discuss that topic in the next two chapters.

## 5.5 Registering the Simple Process: SC_THREAD

Once you have defined a process method, you must identify and register it with the simulation kernel. This step allows the thread to be invoked by the simulation kernel's scheduler. The registration occurs within the module class constructor, **SC_CTOR**, as previously indicated.

Registration of a SystemC thread is coded by using the cpp macro **SC_THREAD** inside the constructor as follows:

```
SC_THREAD(process_name);//Must be INSIDE constructor
```

*Figure 5-7.* Syntax of SC_THREAD

The *process_name* is the name of the corresponding member method of the class. C++ lets the constructor appear before or after declaration of the process method. Here is a complete example of an **SC_THREAD** defined within a module:

```
//FILE: simple_process_ex.h
SC_MODULE(simple_process_ex) {
  SC_CTOR(simple_process_ex) {
    SC_THREAD(my_thread_process);
  }
  void my_thread_process(void);
};
```

*Figure 5-8.* Example of Simple SC_THREAD

---

[19] An **SC_THREAD** is similar to a Verilog **initial** block or a VHDL **process** that ends with a simple **wait;** .

Traditionally, the code above is placed in a header file that has the same name as the module and has a filename extension of .h. Thus, the preceding example could appear inside a file named simple_process_ex.h.

Notice that my_thread_process is not implemented, but only declared. In the manner of C++, it would be legal to implement the member function within the class, but implementations are traditionally placed in a separate file, the .cpp file.

It is also possible to place the implementation of the constructor in the .cpp file, as we shall see in the next section. As an example, the implementation for the my_thread_process would be found in a file named simple_process_ex.cpp, and might contain the following:

```
//FILE: simple_process_ex.cpp
void simple_process_ex::my_thread_process(void) {
  std::cout << "my_thread_process executed within "
          << name()
          << std::endl;
  }
```

*Figure 5-9.* Example of Simple SC_THREAD Implementation

Using **void** inside the declaration parentheses is not required; however, this approach clearly states the intent, and it is a legal construct of C++[20].

Test bench code typically uses **SC_THREAD** processes to accomplish a series of tasks and finally stops the simulation. On the other hand, high-level abstraction hardware models commonly include infinite loops. It is a requirement that such loops explicitly hand over control to other parts of the simulation. This topic will be discussed in Chapter 7, Concurrency.

---

[20] Some situations in C++ using templates require the legality of this syntax.

## 5.6 Completing the Simple Design: main.cpp

Now we complete the design with an example of the top-level file for
`simple_process_ex`. The top-level file for a SystemC model is placed in
the traditional file, `main.cpp`.

```
//FILE: main.cpp
int sc_main(int argc, char* argv[]) { // args unused
  simple_process_ex my_instance("my_instance");
  sc_start();
  return 0; // unconditional success (not
            // recommended)
}
```

*Figure 5-10.* Example of Simple sc_main

Notice the string name constructor argument "`my_instance`" in the
preceding. The reason for this apparent duplication is to store the name of
the instance internally for use when debugging. The **sc_module** class
member function **name**() may be used to obtain the name of the current
instance.

## 5.7 Alternative Constructors: SC_HAS_PROCESS

Before leaving this chapter on modules, we need to discuss an alternative
approach to creating constructors. The alternative approach uses a cpp macro
named **SC_HAS_PROCESS**.

You can use this macro in two situations. First, use **SC_HAS_PROCESS**
when you require constructors with arguments beyond just the instance name
string passed into **SC_CTOR** (e.g., to provide configurable modules). Second,
use **SC_HAS_PROCESS** when you want to place the constructor into the
implementation (i.e., `.cpp` file).

You can use constructor arguments to specify sizes of included
memories, address ranges for decoders, FIFO depths, clock divisors, FFT
depth, and other configuration information. For instance, a memory design
might allow selection of different sizes of memories with an argument:

```
My_memory instance("instance", 1024);
```

*Figure 5-11.* Example of SC_HAS_PROCESS Instantiation

To use this alternative approach, invoke **SC_HAS_PROCESS**, and then create conventional constructors. One caveat applies. You must construct or initialize the module base class, **sc_module**, with an instance name string. That requirement is why the **SC_CTOR** needed an argument. The syntax of this style when used in the header file follows:

```
//FILE: module_name.h
SC_MODULE(module_name) {
  SC_HAS_PROCESS(module_name);
  module_name(sc_module_name instname[, other_args...])
  : sc_module(instname)
  [, other_initializers]
  {
    CONSTRUCTOR_BODY
  }
};
```

*Figure 5-12.* Syntax of SC_HAS_PROCESS in the Header

The syntax for using **SC_HAS_PROCESS** in a separate implementation (i.e., separate compilation situation) is similar.

```
//FILE: module_name.h
SC_MODULE(module_name) {
  SC_HAS_PROCESS(module_name);
  module_name (sc_module_name instname[,other_args...]);
};
```

```
//FILE: module_name.cpp
module_name::module_name(
  sc_module_name instname[, other_args...])
: sc_module(instname)
[, other_initializers]
{
  CONSTRUCTOR_BODY
}
```

*Figure 5-13.* Syntax of SC_HAS_PROCESS Separated

In the preceding examples, the *other_args* are optional.

## 5.8    Two Basic Styles

We finish this chapter with two templates for coding SystemC designs. First, we provide the more traditional style, which leans heavily on headers. Second, our recommended style places more elements into the implementation. Creating a C++ templated module usually precludes this style due to C++ compiler restrictions.

Use either of these templates for your coding and you'll do well. We'll visit these again in more detail in Chapter 10 when we discuss the details of hierarchy and structure.

### 5.8.1 The Traditional Template

The traditional template illustrated in *Figure 5-14* and *Figure 5-15* places all the instance creation and constructor definitions in header (.h) files. Only the implementation of processes and helper functions are deferred to the compiled (.cpp) file. Let's remind ourselves of the basic components in each file. First, the **#ifndef/#define/#endif** prevents problems when the header file is included multiple times. Using NAME_H definition is fairly standard. This definition is followed by file inclusions of any sub-module header files by way of **#include**.

Next, the **SC_MODULE**{…}; surrounds the class definition. Don't forget the trailing semicolon, which is a fairly common error. Within the class definition, ports are usually the first thing declared because they represent the interface to the module. Local channels and sub-module instances come next. We will discuss all of these later in the book.

Next, we place the class constructor, and optionally the destructor. For many cases, the **SC_CTOR**(){…} macro proves quite sufficient for this. The body of the constructor provides initializations, connectivity of sub-modules, and registration of processes. Again, all of this will be discussed in detail in following chapters.

The header finishes out with the declarations of processes, helper functions and possibly other private data. Note that C++ or SystemC does not dictate the ordering of these elements within the class declaration.

```
#ifndef NAME_H
#define NAME_H
#include "submodule.h"
...
SC_MODULE(NAME) {
    Port declarations
    Channel/submodule instances
    SC_CTOR(NAME)
    : Initializations
     {
        Connectivity
        Process registrations
     }
    Process declarations
    Helper declarations
};
#endif
```

*Figure 5-14.* Traditional Style NAME.h Template

The body of a traditional style simply includes the SystemC header file, and the corresponding module header just described. The rest of this file simply contains external function member implementations of the processes and functions, which will be described in upcoming chapters. Note that it is possible to have no implementation file if there are no processes or helper functions in the module.

```
#include <systemc.h>
#include "NAME.h"
NAME::Process {implementations }
NAME::Helper {implementations }
```

*Figure 5-15.* Traditional Style NAME.cpp Template

### 5.8.2 Recommended Alternate Template Form

For various reasons (discussed in Chapter 10, Structure), we recommend a different approach that is more conducive to independent development of modules. For now, we'll just present the template in *Figure 5-16* and *Figure 5-17* and note the differences.

First, the header contains the same #**define** and **SC_MODULE** components as the traditional style. The differences reside in how the channel/sub-module definitions are implemented and movement of the constructor into the implementation body. Notice that the channel/sub-modules are implemented in a different manner (using pointers).

```
#ifndef NAME_H
#define NAME_H
Submodule forward class declarations
SC_MODULE(NAME) {
    Port declarations
    Channel/Submodule* definitions
    // Constructor declaration:
    SC_CTOR(NAME);
    Process declarations
    Helper declarations
};
#endif
```

*Figure 5-16.* Recommended Style NAME.h Template

```
#include <systemc.h>
#include "NAME.h"
 NAME::NAME(sc_module_name nm)
: sc_module(nm)
, Initializations
{
    Channel allocations
    Submodule allocations
    Connectivity
    Process registrations
}
NAME::Process {implementations }
NAME::Helper {implementations }
```

*Figure 5-17.* Recommended Style NAME.cpp Template

## 5.9  Exercises

For the following exercises, use the samples provided at www.EklecticAlly.com/Book/.

**Exercise 5.1:** Compile and run the `simple_process_ex` example from the website. Add an output statement before **`sc_start`**() indicating the end of elaboration and beginning of simulation.

**Exercise 5.2:** Rewrite `simple_process_ex` using **SC_HAS_PROCESS**. Compile and run the code.

**Exercise 5.3:** Add a second **SC_THREAD** to `simple_process_ex`. Be sure the output message is unique. Compile and run.

**Exercise 5.4**: Add a second instantiation of `simple_process_ex`. Compile and run.

**Exercise 5.5**: Write a module from scratch using what you know. The output should count down from 3 to 1 and display the corresponding words "Ready", "Set", "Go" with each count. Compile and run.

Try writing the code without using **SC_MODULE**. What negatives can you think of for not using **SC_MODULE**? [HINT: Think about EDA vendor-supplied tools that augment SystemC.]

| Predefined Primitive Channels: Mutexes, FIFOs, & Signals | | | |
|---|---|---|---|
| **Simulation Kernel** | Threads & Methods | Channels & Interfaces | Data types: Logic, Integers, Fixed point |
| | Events, Sensitivity & Notifications | Modules & Hierarchy | |

Chapter 6

# A NOTION OF TIME

This chapter briefly describes the fundamental notion of time provided by SystemC. We will defer an exploration of many of the intricacies of time until after we discuss events in Chapter 7, Concurrency. Although the time data type itself is simple, the underlying mechanisms are all part of the simulation kernel also discussed in Chapter 7.

## 6.1 sc_time

SystemC provides the **sc_time** data type to measure time. Time is expressed in two parts: a numeric magnitude and a time unit. Possible time unit specifiers are:

```
SC_SEC  // seconds
SC_MS   // milliseconds
SC_US   // microseconds
SC_NS   // nanoseconds
SC_PS   // picoseconds
SC_FS   // femtoseconds
```

*Figure 6-1.* Syntax of sc_time Units

The time data type is declared with the following syntax:

```
sc_time name…; // no initialization
sc_time name(magnitude, timeunits)…;
```

*Figure 6-2.* Syntax of sc_time

SystemC allows addition, subtraction, scaling, and other related operations on **sc_time** objects. Simple examples include:

```
sc_time t_PERIOD(5, SC_NS);
sc_time t_TIMEOUT(100, SC_MS);
sc_time t_MEASURE, t_CURRENT, t_LAST_CLOCK;
t_MEASURE = (t_CURRENT-t_LAST_CLOCK);
if (t_MEASURE > t_HOLD) { error("Setup violated") }
```

*Figure 6-3.* Examples of sc_time

Note the convention used to identify time variables. This convention aids understanding of code. In addition, hard coding of constants is discouraged because it reduces both readability and reusability. One special constant should be noted, **SC_ZERO_TIME**, which is simply **sc_time(0,SC_SEC)**.

## 6.2  sc_start()

**sc_start()** is a key method in SystemC. This method starts the simulation phase, which consists of initialization and execution. Of interest to this chapter, **sc_start()** takes an optional argument of type **sc_time**. This syntax lets you specify a maximum simulation time. Without an argument, **sc_start()** specifies that the simulation may run forever. If you provide a time argument, simulation stops after the specified simulation time has elapsed.

```
sc_start();              //sim "forever"
sc_start(max_sc_time);//sim no more than max_sc_time
```

*Figure 6-4.* Syntax of sc_start()

The following example, which is based on the previous chapter's simple process example, illustrates limiting the simulation to sixty seconds.

```
int sc_main(int argc, char* argv[]) { // args unused
  simple_process_ex my_instance("my_instance");
  sc_start(60.0,SC_SEC); // Limit sim to one minute
  return 0;
}
```

*Figure 6-5.* Example of sc_start()

Note that internally, SystemC represents time with a 64-bit integer (**uint64**). This data type can represent a very long time but not infinite time.

## 6.3 sc_time_stamp () and Time Display

SystemC's simulation kernel keeps track of the current time and is accessible with a call to the **sc_time_stamp**() method.

```
cout << sc_time_stamp() << endl;
```

*Figure 6-6.* Example of sc_time_stamp()

Additionally, the **ostream operator<<** has been overloaded to allow convenient display of time.

```
ostream_object << desired_sc_time_object;
```

*Figure 6-7.* Syntax of ostream << overload

Here is a simple example and corresponding output:

```
std::cout << "  The time is now "
          << sc_time_stamp()
          << "!" << std::endl;
```

*Figure 6-8.* Example of sc_time_stamp () and ostream << overload

```
The time is now 0 ns!
```

*Figure 6-9.* Output of sc_time_stamp() and ostream << overload

A more complete example follows in the next section.

## 6.4  wait(sc_time)

It is often useful to delay a process for specified periods of time. You can use this delay to simulate delays of real activities (e.g., mechanical actions, chemical reaction times, or signal propagation). The **wait**() method provides syntax allowing delay specifications in **SC_THREAD** processes. When a **wait**() is invoked, the **SC_THREAD** process blocks itself and is resumed by the scheduler at the specified time. We will discuss **SC_THREAD** processes in further detail in the next chapter.

```
wait(delay_sc_time); // wait specified amount of time
```

*Figure* 6-10. Syntax of wait () with a Timed Delay

Here are some simple examples:

```
void simple_process_ex::my_thread_process(void) {
  wait(10,SC_NS);
  std::cout<< "Now at "<< sc_time_stamp()<< std::endl;
  sc_time t_DELAY(2,SC_MS); // keyboard debounce time
  t_DELAY *= 2;
  std::cout<< "Delaying "<< t_DELAY<< std::endl;
  wait(t_DELAY);
  std::cout << "Now at " << sc_time_stamp()
            << std::endl;
}
```

```
% ./run_example
Now at 10 ns
Delaying 4 ms
Now at 4000010 ns
```

*Figure* 6-11. Example of wait()

## 6.5  sc_simulation_time(), Time Resolution and Time Units

There are times when you may want to manipulate time in a native C++ data type and shed the time units. For this purpose, **sc_simulation_time**() returns time as a **double** in the current default time unit.

```
sc_simulation_time()
```

*Figure* 6-12. Syntax of sc_simulation_time ()

To establish the default time unit, call **sc_set_default_time_unit**(). You must call this routine prior to all time specifications and prior to the initiation of **sc_start**(). You may precede this call by specifying the time resolution using **sc_set_time_resolution**().

These methods have the following syntax:

```
//positive power of ten for resolution
sc_set_time_resolution(value, tunit);
//power of ten >= resolution for default time unit
sc_set_default_time_unit(value, tunit);
```

*Figure* 6-13. Syntax to set time units and resolution

Rounding will occur if you specify a time constant within the code that has a resolution more precise than the resolution specified by this routine. For example, if the specified time resolution is 100 ps, then coding 20 ps will result in an effective value of 0 ps.

Because the simulator has to keep enough information to represent time to the specified resolution, the time resolution can also have an effect on simulation performance. This result depends on the simulation kernel implementation.

GUIDELINE:   Do not specify more resolution than the design needs.

The following example uses the **sc_time** data type and several of the methods discussed in this chapter. For this example, we know that the simulation will not run for more than two hours (or 7200 seconds). The value of t will be between 0.000 and 7200.000 since the resolution is in milliseconds.

```
int sc_main(int argc, char* argv[]) {// args unused
  sc_set_time_resolution(1,SC_MS);
  sc_set_default_time_unit(1,SC_SEC);
  simple_process_ex my_instance("my_instance");
  sc_start(7200,SC_SEC); // Limit simulation to 2
                         // hours
  double t = sc_simulation_time();
  unsigned hours   = int(t / 3600.0);
  t -= 3600.0*hours;
  unsigned minutes = int(t / 60.0);
  t -= 60.0*minutes;
  double   seconds = t;
  cout<< hours<< " hours "
      << minutes<< " minutes "
      << seconds<< " seconds" //to the nearest ms
      << endl;
  return 0;
}
```

*Figure* 6-14. Example of sc_time Data Type

## 6.6　Exercises

For the following exercises, use the samples provided in www.EklecticAlly.com/Book/.

**Exercise 6.1:** Examine, compile, and run the example `time_flies`, found on the website.

**Exercise 6.2:** Modify `time_flies` to see how much time you can model (days? months?). See how it changes with the time resolution.

**Exercise 6.3:** Copy the basic structure of `time_flies` and model one cylinder of a simple combustion engine. Modify the body of the thread function to represent physical actions using simple delays. Use **std::cout** statements to indicate progress.

Suggested activities include opening the intake, adding fuel and air, closing the intake, compressing gas, applying voltage to the spark plug, igniting fuel, expanding gas, opening the exhaust valves, closing the exhaust valves. Use delays representative of 800 RPM. Use time variables with appropriate names. Compile and run.

# Chapter 7

# CONCURRENCY
*Processes & Events*

Many activities in a real system occur at the same time or concurrently. For example, when simulating something like a traffic pattern with multiple cars, the goal is to model the vehicles independently. In other words, the cars operate in parallel.

Software typically executes using a single thread of activity, because there is usually only one processor on which to run, and partly because it is much easier to manage. On the other hand, in real systems many things occur simultaneously. For example, when an automobile executes a left turn, it is likely that the left turn indicator is flashing, the brakes are engaged to slow down the vehicle, engine power is decreased as the driver lets off the accelerator, and the transmission is shifting to a lower gear. All of these activities can occur at the same instant.

SystemC uses processes to model concurrency. As with most event-driven simulators, concurrency is not true concurrent execution. In fact, simulated concurrency works like cooperative multi-tasking. In other words, the concurrency is not pre-emptive. Each process in the simulator executes a small chunk of code, and then voluntarily releases control to let other processes execute in the same simulated time space.

As with any simulator, understanding how the simulator handles concurrency enables the designer to use the simulation kernel more effectively and write more efficient models. This understanding will also help avoid many traps.

The simulator kernel is responsible for starting processes and managing which process executes next. Due to the cooperative nature of the simulator model, processes are responsible for suspending themselves to allow execution of other concurrent processes.

SystemC presently provides two major process types, **SC_THREAD** processes and **SC_METHOD** processes. A third type, the **SC_CTHREAD**, is a minor variation on the **SC_THREAD** process that we will discuss separately as an advanced topic in Chapter 14.

## 7.1  sc_event

Before we can discuss how processes work in the simulator, it is necessary to discuss events. Events are key to an event-driven simulator like the SystemC simulation kernel.

An event is something that happens at a specific point in time. An event has no value and no duration. SystemC uses the **sc_event** class to model events. This class allows explicit launching or triggering of events by means of a notification method.

DEFINITION: A SystemC event is the occurrence of an **sc_event** notification and happens at a single point in time. An event has no duration or value.

Once an event occurs, there is no trace of its occurrence other than the side effects that may be observed as a result of processes that were sensitive to or waiting for the event. The following diagram illustrates an event e_rdy "firing" at three different points. Note that unlike a waveform, events have no time width.

**Event
Timeline**

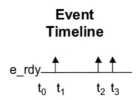

*Figure 7-1.* Graphical Representation of an Event

Because events have no duration, you must be watching to catch them. Quite a few coding errors are due to not understanding this simple rule. Let's restate it.

RULE:      To observe an event, the observer must be watching for the event.

SystemC lets processes wait for an event by using a dynamic or static sensitivity that we will discuss shortly. If an event occurs, and no processes are waiting to catch it, the event goes unnoticed. The syntax to declare a named event is simple:

```
sc_event event_name₁[,event_nameᵢ]…;
```

*Figure 7-2.* Syntax of sc_event

Remember that **sc_event**s have no value, and you can perform only two actions with an **sc_event**: wait for it or cause it to occur. We will discuss details in the next sections.

## 7.2  Simplified Simulation Engine

Before proceeding, we need to understand how event-driven simulation works. The following simplified[21] flow diagram illustrates the operation of the SystemC simulation kernel.

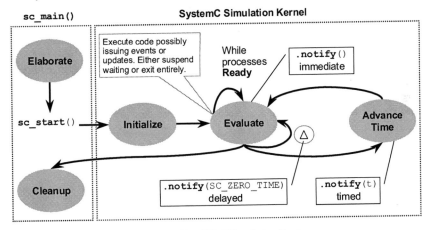

*Figure 7-3.* Simplified Simulation Engine

First, elaboration occurs as previously discussed in Chapter 5, Modules. During elaboration, SystemC modules are constructed and various simulation parameters are established. This elaboration phase is followed by a call to **sc_start**(), which invokes the simulation kernel. This call begins the initialization phase. Processes (e.g., **SC_THREAD** processes) defined during elaboration need to be started. During the initialization phase, all[22] processes are placed initially into a ready pool.

DEFINITION:  A process is ready whenever it is available to be executed.

---

[21] We discuss the remaining details in Chapter 9, Evaluate-Update Channels.
[22]  Actually most, but we will discuss this in a later section of this chapter.

We sometimes say that processes are placed into the pool of processes ready to execute. The following diagram depicts process and event pools.

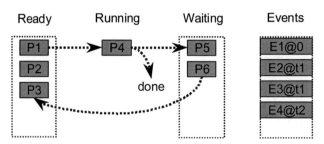

*Figure 7-4* Process and Event Pools

Simulation then proceeds into the evaluation phase. One by one processes are randomly[23] taken from the ready pool by designating them as running and invoked. Each process executes until it either completes (e.g., via a **return**) or suspends (e.g., calls **wait()**).

During execution, a process may invoke immediate event notification (i.e., *event*.**notify()**) and possibly cause one or more waiting processes to be placed in the ready state. It is also possible to generate delayed or timed event notifications as indicated by E1–E4 in the preceding figure. Completed processes are discarded. Suspended processes are placed into a waiting pool. Simulation proceeds until there are no more processes ready to run.

DEFINITION:  SystemC enters the waiting state whenever it encounters an explicit **wait()**, or in some cases performs a **return**.

At this point, execution exits the evaluate bubble at the bottom of *Figure 7-3* with one of three possibilities: waiting processes or events that are zero time delayed, non-zero time delayed, or neither.

First, there may be processes or events (to be discussed shortly) waiting for an **SC_ZERO_TIME** delay (.**notify**(0)). This delay is known as a delta-cycle delay. In this case, waiting pool processes with zero time delays are placed back into the ready pool. Zero time events originating from delayed notifications may cause processes waiting on those events to also be placed into the ready pool. Another round of evaluation occurs if any processes have been moved into the ready pool.

---

[23] Although not truly random, the SystemC specification allows that different implementations may choose any ordering as is convenient for simulation.

Second, there may be processes or events, scheduled for later, waiting for a non-zero time delay to occur. In this case, time is advanced to the nearest time indicated. Processes waiting on that specific delay will be placed into the ready pool. If an event occurs at this new time, processes waiting on that event are placed into the ready pool. Another round of evaluation occurs if any processes have been moved into the ready pool.

Third, it is possible that there were no delayed processes or events. Since there are no processes in the ready pool, then the simulation simply ends by returning to **sc_main** and finishing cleanup. It is not possible to successfully re-enter **sc_start** in the current definition of SystemC.

## 7.3 SC_THREAD

**SC_THREAD** processes are started once and only once by the simulator. Once a thread starts to execute, it is in complete control of the simulation until it chooses to return control to the simulator.

SystemC offers two ways to pass control back to the simulator. One way is to simply exit (e.g., **return**), which has the effect of terminating the thread for the rest of the simulation. When an **SC_THREAD** process exits, it is gone forever, therefore **SC_THREAD**s typically contain an infinite loop containing at least one **wait**.

The other way to return control to the simulator is to invoke the module **wait** method. The **wait** suspends the **SC_THREAD** process.

Sometimes **wait** is invoked indirectly. For instance, a blocking **read** or **write** of the **sc_fifo** invokes **wait** when the FIFO is empty or full, respectively. In this case, the **SC_THREAD** process suspends similarly to invoking **wait** directly.

## 7.4 Dynamic Sensitivity for SC_THREAD::wait()

As indicated previously, **SC_THREAD** processes rely on the **wait** method to suspend their execution. The **wait** method supplied by the **sc_module** class has several syntaxes as indicated below. When **wait** executes, the state of the current thread is saved, the simulation kernel is put in control and proceeds to activate another ready process. When the suspended process is reactivated, the scheduler restores the calling context of the original thread, and the process resumes execution at the statement after the **wait**.[24]

```
wait(time);
wait(event);
wait(event₁ | eventₙ...); // any of these
wait(event₁ & eventₙ...); // all of these
wait(timeout, event);    // event with timeout
wait(timeout, event₁ | eventₙ...);//any event with
                                 // timeout
wait(timeout, event₁ & eventₙ...);//all events with
                                 // timeout
wait(); // static sensitivity
```

*Figure 7-5.* Syntax of SC_THREAD wait()

We have already described the first syntax in Chapter 6, A Notion of Time; this syntax provides a delay for a specified period. The next several forms specify events and suspend execution until one or all the events have occurred. The operator | is defined to mean any of these events; whichever one happens first will cause a return to **wait**. The operator & is defined to mean all of these events in any order must occur before **wait** returns. The last syntax, **wait()**, will be deferred to a joint discussion with static sensitivity later in this chapter.

Use of a *timeout* is handy when testing protocols and various error conditions. When using a *timeout* syntax, the Boolean function **timed_out()** may be called immediately after the **wait** to determine whether a time out caused execution to resume.

```
...
sc_event ack_event, bus_error_event;
...
wait(t_MAX_DELAY, ack_event | bus_error_event);
if (timed_out()) break; // path for a time out
    ...
```

*Figure 7-6.* Example of timed_out() and wait()

---

[24] This magic is handled by the SystemC library's scheduler. A detailed description of how context switches are managed goes beyond the scope of this book.

Notice when multiple events are or'ed, it is not possible to know which event occurred in a multiple event wait situation as events have no value (Section 7.1). Thus, it is illegal to test an event for **true** or **false**.

```
if (ack_event) do_something; // syntax error!
```

*Figure 7-7.* Example of Illegal Boolean Compare of sc_event()

It is ok to test a Boolean that is set by the process that caused an event; however, it is problematic to clear it properly. We now need to learn how to generate event occurrences.

## 7.5   Another Look at Concurrency and Time

Let's take a look at a hypothetical example to better understand how time and execution interact. Consider a design with four processes as illustrated in *Figure 7-8*. For this discussion, assume the times, $t_1$, $t_2$, and $t_3$ are non-zero. Each process contains lines of code or statements ($stmt_{A1}$, $stmt_{A2}$, ...) and executions of wait methods (**wait**(t1), **wait**(t2),...)

```
Process_A() {           Process_B() {           Process_C() {           Process_D() {
  //@ t0                  //@ t0                  //@ t0                  //@ t0
  stmt_A1;                stmt_B1;                stmt_C1;                stmt_D1;
  stmt_A2;                stmt_B2;                stmt_C2;                stmt_D2;
  wait(t1);              wait(t1);              wait(t1);              wait(t1);
  stmt_A3;                stmt_B3;                stmt_C3;                stmt_D3;
  stmt_A4;                stmt_B4;                stmt_C4;                wait(
  wait(t2);a            wait(t2);              wait(t2);                SC_ZERO_TIME);
  stmt_A5;                stmt_B5;                stmt_C5;                stmt_D4;
  stmt_A6;                stmt_B6;                stmt_C6;                wait(t3);
  wait(t3);              wait(t3);              wait(t3);              }
}                       }                       }
```

*Figure 7-8.* Four Processes

Notice that Process_D skips $t_2$. At first glance, it might be perceived that time passes as shown below. The uninterrupted solid and hatched line portions indicate program activity. Vertical discontinuities indicate a wait.

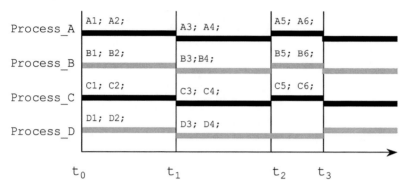

*Figure 7-9.* Simulated Activity—Perceived

Each process' statements take some time and are evenly distributed along simulation time. Perhaps surprisingly that is <u>not</u> how it works at all. In fact, actual simulated activity is shown in the next figure.

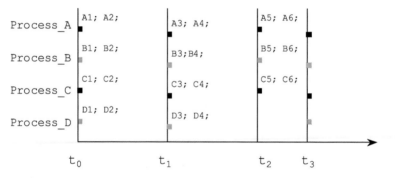

*Figure 7-10.* Simulated Activity—Actual

Each set of statements executes in zero time! Let's expand the time scale to expose the simulator's time as well. This expansion exposes the operation of the scheduler at work.

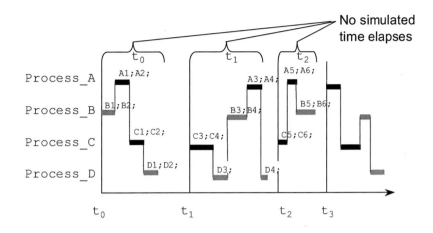

*Figure 7-11.* Simulated Activity with Simulator Time Expanded

Notice that the ordering of processes appears quite non-deterministic in this example. This non-determinism is specified by the SystemC standard. Now for any given simulator and set of code, there is also a requirement that the simulation be deterministic in the sense that one may rerun the simulation and obtain the same results.

All of the executed statements in this example execute during the same evaluate phase of a delta cycle. If any of the statements had been a delayed notification, then multiple delta cycles may have occurred during the same instant in time.

As a final consideration, the previous diagrams would be equally valid with any or all of the indicated times $t_1$, $t_2$, or $t_3$ as zero (i.e., **SC_ZERO_TIME**). Once you grasp these fundamental concepts, understanding SystemC behaviors will become much easier.

## 7.6   Triggering Events: .notify()

Events occur explicitly by using the **notify**() method. This method has two syntax styles. The authors prefer the object-oriented style to the function-call style.

```
// Object-oriented style (preferred)
event_name.notify(); //immediate notification
event_name.notify(SC_ZERO_TIME);// delayed
                              // notification
event_name.notify(time); //timed notification

// Functional-call style
notify(event_name); //immediate notification
notify(event_name, SC_ZERO_TIME);// delayed
                              // notification
notify(event_name, time); //timed notification
```

*Figure 7-12.* Syntax of notify()

Invoking an immediate **notify**() causes any processes waiting for the event to be immediately moved from the wait pool into the ready pool for execution.

Delayed notification occurs when a time of zero is specified. Processes waiting for a delayed notification will execute only after all waiting processes have executed or in other words executes on the next delta-cycle (after an update phase). This is quite useful as we shall see shortly. There is an alternate syntax of **.notify_delayed**(), but this syntax may be deprecated since it is redundant.

Timed notification would appear to be the easiest to understand. Timed events are scheduled to occur at some time in the future.

One confounding aspect of timed events, which includes delayed events, concerns multiple notifications. An **sc_event** may have no more than a single outstanding scheduled event, and only the nearest time notification is allowed. If a nearer notification is scheduled, the previous outstanding scheduled event is canceled.

In fact, scheduled events may be canceled with the **.cancel**() method. Note that immediate events cannot be canceled because they happen at the precise moment they are notified (i.e., immediately).

```
event_name.cancel();
```

*Figure 7-13.* Syntax of cancel() Method

The best way to understand events is by way of examples. Notice in the following example that all of the **notify**s execute at the same instant in time.

```
...
sc_event action;
sc_time now(sc_time_stamp()); //observe current time
//immediately cause action to fire
action.notify();
//schedule new action for 20 ms from now
action.notify(20,SC_MS);
//reschedule action for 1.5 ns from now
action.notify(1.5,SC_NS);
//useless, redundant
action.notify(1.5,SC_NS);
//useless preempted by event at 1.5 ns
action.notify(3.0,SC_NS);
//reschedule action for next delta cycle
action.notify(SC_ZERO_TIME);
//useless, preempted by action event at SC_ZERO_TIME
action.notify(1,SC_SEC);
//cancel action entirely
action.cancel();
//schedule new action for 1 femtosecond from now
action.notify(20,SC_FS);
...
```

*Figure 7-14.* Example of sc_event notify() and cancel() Methods

To illustrate the use of events, let's consider how one might model the interaction between switches on a steering wheel column and the remotely-located signal indicators (lamps). The following example models a mechanism that detects switching activity and notifies the appropriate indicator. For simplicity, only the stoplight interaction is modeled here.

In this model, the process turn_knob_thread provides a stimulus and interacts with the process stop_signal_thread. The idea is to have several threads representing different signal indicators, and turn_knob_thread directs each indicator to turn on or off via the signal_stop and signals_off events. The indicators provide their status via the stop_indicator_on and stop_indicator_off events.

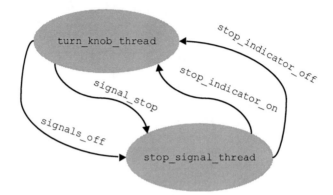

*Figure 7-15.* Turn of Events Illustration

```
//FILE: turn_of_events.h
SC_MODULE(turn_of_events) {
  // Constructor
  SC_CTOR(turn_of_events) {
    SC_THREAD(turn_knob_thread);
    SC_THREAD(stop_signal_thread);
  }
  sc_event signal_stop, signals_off;
  sc_event stop_indicator_on, stop_indicator_off;
  void turn_knob_thread(); // stimulus process
  void stop_signal_thread(); // indicator process
};//endclass turn_of_events
```

*Figure 7-16.* Example of turn_of_events Header

An interesting aspect of the implementation shown in the following figure is consideration of process ordering effects. Recall the rule that "To see an event, a process must be waiting for it." It is because of this requirement that the turn_knob_thread implementation starts out with **wait(SC_ZERO_TIME)**. Without that pause, if turn_knob_thread runs first, then the stop_signal_thread will never see any events because it will not have executed the first **wait()**. As a result, the simulation would starve and exit.

Similarly, consider what would happen if the signals_off event were issued before signal_stop. If an unconditional wait for acknowledgement

occurred, the simulation would exit. It would exit because the
turn_knob_thread would be waiting on an event that never occurs
because the stop_signal_thread was not in a position to issue that
event.

```cpp
//FILE: turn_of_events.cpp
void turn_of_events::turn_knob_thread() {
  // This process provides stimulus using stdin
  enum directions {STOP="S", OFF="F"};
  char direction; // Selects appropriate indicator
  bool did_stop = false;
  // allow other threads to get into waiting state
  wait(SC_ZERO_TIME);
  for(;;) {
    // Sit in an infinite loop awaiting keyboard
    // or STDIN input to drive the stimulus...
    std::cout << "Signal command: ";
    std::cin >> direction;
    switch (direction) {
      case STOP:
        // Make sure the other signals are off
        signals_off.notify();
        signal_stop.notify(); // Turn stop light on
        // Wait for acknowledgement of indicator
        wait(stop_indicator_on);
        did_stop = true;
        break;
      case OFF:
        // Make the other signals are off
        signals_off.notify();
        if (did_stop) wait(stop_indicator_off);
        did_stop = false;
        break;
    }//endswitch
  }//endforever
}//end turn_knob_thread()
```

*Figure 7-17.* Example of Turn of Events Stimulus

```
void turn_of_events::stop_signal_thread() {
  for(;;) {
    wait(signal_stop);
    std::cout << "STOPPING      !!!!!!" << std::endl;
    stop_indicator_on.notify();
    wait(signals_off);
    std::cout << "Stop off      ------" << std::endl;
    stop_indicator_off.notify();
  }//endforever
}//end stop_signal_thread()
```

*Figure 7-18.* Example of Turn of Events Indicator

The preceding example, turn_of_events, models two processes with SystemC threads. The turn_knob_thread takes input from the keyboard and notifies the stop_signal_thread. Sample output might look as follows (user input highlighted):

```
% ./turn_of_events.x
Signal command: S
STOPPING  !!!!!!
Signal command: F
Stop off  ------…
```

*Figure 7-19.* Example of Turn of Events Output

## 7.7 SC_METHOD

As mentioned earlier, SystemC has more than one type of process. The **SC_METHOD** process is in some ways simpler than the **SC_THREAD**; however, this simplicity makes it more difficult to use for some modeling styles. To its credit, **SC_METHOD** is more efficient than **SC_THREAD**.

What is different about an **SC_METHOD**? One major difference is invocation. **SC_METHOD** processes never suspend internally (i.e., they can never invoke **wait()**). Instead, **SC_METHOD** processes run completely and return. The simulation engine calls them repeatedly based on the dynamic or static sensitivity. We will discuss both shortly.

In some sense, **SC_METHOD** processes are similar to the Verilog **always@** block or the VHDL **process**. By contrast, if an **SC_THREAD** terminates, it never runs again in the current simulation.

Because **SC_METHOD** processes are prohibited from suspending internally, they may not call the **wait** method. Attempting to call **wait** either directly or implied from an **SC_METHOD** results in a runtime error.

Implied waits result from calling SystemC built-in methods that are defined such that they may issue a **wait**. These are known as blocking methods. The **read** and **write** methods of the **sc_fifo** data type, discussed later in this book, are examples of blocking methods. Thus, **SC_METHOD** processes must avoid using calls to blocking methods.

The syntax for **SC_METHOD** processes follows and is almost identical to **SC_THREAD** except for the keyword **SC_METHOD**:

```
SC_METHOD(process_name);//Located INSIDE constructor
```

*Figure 7-20.* Syntax of SC_METHOD

A note on the choice of these keywords might be useful. The similarity of name between an **SC_METHOD** process and a regular object-oriented method betrays its name. It executes without interruption and returns to the caller (the scheduler). By contrast, an **SC_THREAD** process is more akin to a separate operating system thread with the possibility of being interrupted and resumed.

Variables allocated in **SC_THREAD** processes are persistent. **SC_METHOD** processes must declare and initialize variables each time the method is invoked. For this reason, **SC_METHOD** processes typically rely on module local data members declared within the **SC_MODULE**. **SC_THREAD** processes tend to use locally declared variables.

GUIDELINE: To differentiate threads from methods, we strongly recommend adopting a naming style. One naming style appends _thread or _method as appropriate. Being able to differentiate processes based on names becomes useful during debug.

## 7.8   Dynamic Sensitivity for SC_METHOD: next_trigger()

**SC_METHOD** processes dynamically specify their sensitivity by means of the **next_trigger**() method. This method has the same syntax as the **wait**() method but with a slightly different behavior.

```
next_trigger(time);
next_trigger(event);
next_trigger(event₁ | eventᵢ…); //any of these
next_trigger(event₁ & eventᵢ…); //all of these
                                //required
next_trigger(timeout, event);   //event with timeout
next_trigger(timeout, event₁ | eventᵢ…);//any + timeout
next_trigger(timeout, event₁ & eventᵢ…);//all + timeout
next_trigger(); //re-establish static sensitivity
```

*Figure 7-21.* Syntax of SC_METHOD next_trigger()

As with **wait**, the multiple event syntaxes do not specify order. Thus, with **next_trigger**(evt1 & evt2), it is not possible to know which occurred first. It is only possible to assert that both evt1 and evt2 happened.

The **wait** method suspends **SC_THREAD** processes; however, **SC_METHOD** processes are not allowed to suspend. The **next_trigger** method has the effect of temporarily setting a sensitivity list that affects the **SC_METHOD**. **next_trigger** may be called repeatedly, and each invocation encountered overrides the previous. The last **next_trigger** executed before a return from the process determines the sensitivity for a recall of the process. The initialization call is vital to making this work. See the **next_trigger** code in the downloads section of the website for an example.

You should note that it is critical for EVERY path through an **SC_METHOD** to specify at least one **next_trigger** for the process to be called by the scheduler. Without a **next_trigger** or static sensitivity (discussed in the next section), an **SC_METHOD** will never be executed again. A safeguard can be adopted of placing a default **next_trigger** as the first statement of the **SC_METHOD**, since subsequent **next_triggers** will overwrite any previous. A better way to manage this problem exists as we will now discuss.

## 7.9   Static Sensitivity for Processes

Thus far, we've discussed techniques of dynamically (i.e., during simulation) specifying how a process will resume (either by **SC_THREAD** using **wait** or by **SC_METHOD** using **next_trigger**). SystemC provides another type of sensitivity for processes called static sensitivity. Static sensitivity establishes the parameters for resuming during elaboration (i.e., before simulation begins). Once established, static sensitivity parameters cannot be changed (i.e. they're static). It is possible to override static sensitivity as we'll see.

Static sensitivity is established with a call to the **sensitive**() method or the overloaded stream **operator<<** that is placed just following the registration of a process. Static sensitivity applies only to the most recent process registration. **sensitive** may be specified repeatedly. There are two syntax styles:

```
// IMPORTANT: Must follow process registration
sensitive << event [<< event]…; // streaming style
sensitive(event [, event]…);    // functional style
```

*Figure 7-22.* Syntax of sensitive

We prefer the streaming style as it feels more object-oriented, reduces typing, and keeps us focused on C++.

For the next few sections, we will examine the problem of modeling access to a gas station to illustrate the use of sensitivity coupled with events. Initially, we model a single pump station with an attendant and only two customers. The declarations for this example in *Figure 7-24* illustrate the use of the **sensitive** method.

## Early Gas Station

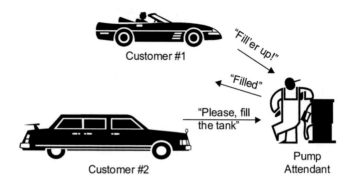

*Figure 7-23.* Initial Gas Station Illustration

```
SC_MODULE(gas_station) {
  sc_event e_request1, e_request2;
  sc_event e_tank_filled;
  SC_CTOR(gas_station) {
    SC_THREAD(customer1_thread);
      sensitive(e_tank_filled); // functional
                                // notation
    SC_METHOD(attendant_method);
      sensitive << e_request1
                << e_request2; // streaming notation
    SC_THREAD(customer2_thread);
  }
  void attendant_method();
  void customer1_thread();
  void customer2_thread();
};
```

*Figure 7-24.* Example of Gas Station Declarations

The gas_station module has two processes with different sensitivity lists and one, customer2_thread, which has none. The attendant_method implicitly executes every time an e_request1 or e_request2 event occurs (unless dynamic sensitivity is invoked by the simulation process).

Notice the indentation used in *Figure 7-24*. This format helps draw attention to the sensitivity being associated with only the most recent process registration.

Here are some fragments of the implementation code focused on the elements of this chapter. You can find the full code in the downloads section of the website.

```cpp
...
void gas_station::customer1_thread() {
  for (;;) {
    wait(EMPTY_TIME);
    cout << "Customer1 needs gas" << endl;
    do {
      e_request1.notify();
      wait(); // use static sensitivity
    } while (m_tank1 == 0);
  }//endforever
}//end customer1_thread()

// omitting customer2_thread (almost identical
// except using wait(e_request2);)

void gas_station::attendant_method() {
  if (!m_filling) {
    ...
    cout << "Filling tank" << endl;
    m_filling = true;
    next_trigger(FILL_TIME);
    ...
  } else {
    ...
    e_filled.notify(SC_ZERO_TIME);
    cout << "Filled tank" << endl;
    ...
    m_filling = false;
    ...
  }//endif
}//end attendant_method()
```

*Figure 7-25.* Example of Gas Station Implementation

The preceding code produces the following output:

```
...
Customer1 needs gas
Filling tank
Filled tank
...
```

*Figure 7-26.* Example of Gas Station Sample Output

## 7.10 dont_initialize

The simulation engine description specifies that all processes are executed initially. This approach makes no sense in the preceding gas_station model as the attendant_method would fill the tank before being requested.

Thus, it becomes necessary to specify some processes that are not initialized. For this situation, SystemC provides the **dont_initialize** method. The syntax follows:

```
// IMPORTANT: Must follow process registration
dont_initialize();
```

*Figure 7-27.* Syntax of dont_initialize()

Note that the use of **dont_initialize** requires a static sensitivity list; otherwise, there would be nothing to start the process. Now our gas_station module contains:

```
    ...
    SC_METHOD(attendant_method);
      sensitive(fillup_request);
      dont_initialize();
    ...
```

*Figure 7-28.* Example of dont_initialize()

## 7.11 sc_event_queue

In light of the limitation that **sc_event**s may only have a single outstanding schedule, **sc_event_queue**s have been added in SystemC version 2.1. These additions let a single event be scheduled multiple times even for the same time! When events are scheduled for the same time, each happens in a separate delta cycle.

```
sc_event_queue event_name₁("event_name₁") …;
```

*Figure 7-29.* Syntax of sc_event_queue

**sc_event_queue** is slightly different from **sc_event**. First, **sc_event_queue** objects do not support immediate notification since there is obviously no need to queue these. Second, the **.cancel**() method is replaced with **.cancel_all**() to emphasize that it cancels all outstanding **sc_event_queue** notifications.

```
...
sc_event_queue action;
sc_time now(sc_time_stamp()); //observe current time
action.notify(20,SC_MS);//schedule for 20 ms from now
action.notify(1.5,SC_NS);//another for 1.5 ns from
                         //now
action.notify(1.5,SC_NS);//another identical action
action.notify(3.0,SC_NS);//another for 3.0 ns from
                         //now
action.notify(SC_ZERO_TIME);//for next delta cycle
action.notify(1,SC_SEC); //for 1 sec from now
action.cancel_all(); // cancel all actions entirely
...
```

*Figure 7-30.* Example of sc_event_queue

The **.cancel**() method is not currently implemented; although, an obvious extension might be to allow canceling notifications at specific times. Another extension might be obtaining information on how many outstanding notifications exist (**.pending**()).

## 7.12 Exercises

For the following exercises, use the samples provided in www.EklecticAlly.com/Book/.

**Exercise 7.1**: Examine, predict the behavior, compile, and run the `turn_of_events` example.

**Exercise 7.2**: Examine, predict the behavior, compile, and run the `gas_station` example.

**Exercise 7.3:** Examine, predict the behavior, compile, and run the `method_delay` example.

**Exercise 7.4**: Examine, predict the behavior, compile, and run the `next_trigger` example.

**Exercise 7.5**: Examine, predict the behavior, compile, and run the `event_filled` example. If using SystemC version 2.1, compile the simulation a second time with the macro `SYSTEMC21` defined (-`DSYSTEMC21` command-line option for gcc).

# Chapter 8

# BASIC CHANNELS

Thus far, we have communicated information between concurrent processes using events and using ordinary module member data. Because there are no guarantees about which processes execute next from the ready state, we must be extremely careful when sharing data.

Events let us manage this aspect of SystemC, but they require careful coding. Because events may be missed, it is important to update a handshake variable indicating when a request is made, and clear it when the request is acknowledged.

## Data Communication

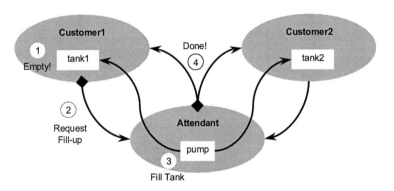

*Figure 8-1.* Gas Station Processes and Events

Let's consider the gas station example again as illustrated in *Figure 8-1*. The customer notices an empty tank ①. The attendant has to be watching when the customer requests a fill-up ②, and make note of it if in the middle of filling up another customer ③. More to the point, if the attendant waits on either customer's request (i.e., **wait**(e_request1|e_request2)), the semantics of **sc_event** doesn't allow the attendant to know which customer made the request after the event happens. That is why the `gas_station` model uses the status of the gas tank as an indicator to choose whether to fill the tank. Similarly, the customer must watch to see if the tank actually was filled when the attendant yells done ④.

SystemC has built-in mechanisms, known as channels, to reduce the tedium of these chores, aid communications, and encapsulate complex communications. SystemC has two types of channels: primitive and hierarchical. This chapter concerns itself with primitive channels. Hierarchical channels are the subject matter of Chapter 13, Custom Channels.

## 8.1  Primitive Channels

SystemC's primitive channels are known as primitive because they contain no hierarchy, no processes, and are designed to be very fast due to their simplicity. All primitive channels inherit from the base class **`sc_prim_channel`**.

SystemC contains several built-in primitive channels. We will discuss channels exhibiting evaluate-update behavior in Chapter 9, Evaluate-Update Channels. In Chapter 13, Custom Channels, we will discuss custom primitive channels as an advanced topic. This chapter focuses on the simplest channels, **`sc_mutex`**, **`sc_semaphore`**, and **`sc_fifo`**.

## 8.2  sc_mutex

Mutex is short for mutual exclusion object. In computer programming, a mutex is a program object that lets multiple program threads share a common resource, such as file access, without colliding.

During elaboration, a mutex is created with a unique name; subsequently, any process that needs the resource must lock the mutex to prevent other processes from using the shared resource. The process should unlock the mutex when the resource is no longer needed. If another process attempts to access a locked mutex, that process is prevented until the mutex becomes available (unlocked).

SystemC provides mutex via the **`sc_mutex`** channel. This class contains several access methods including both blocked and unblocked styles. Remember blocking methods can only be used in **`SC_THREAD`** processes.

```
sc_mutex NAME;

// To lock the mutex NAME (wait until
// unlocked if in use)
NAME.lock();

// Non-blocking, returns true if success, else false
NAME.trylock()

// To free a previously locked mutex
NAME.unlock();
```

*Figure 8-2.* Syntax of sc_mutex

The example of the gas station attendant is a good example of a resource that needs to be shared. Only one car at a time is filled, assuming there is only a single gas pump.

Another example of a resource requiring a mutex would be the controls of an automobile. Only one driver at a time can sit in the driver's seat. In a simulation modeling the interaction of drivers across town with a variety of vehicles, this might be interesting to model.

```
sc_mutex drivers_seat;
...
car->drivers_seat.lock(); // sim driver acquiring
                          // driver's seat
car->start();
... // operate vehicle
car->stop();
car->drivers_seat.unlock(); // sim driver leaving
                           // vehicle
```

*Figure 8-3.* Example of sc_mutex

An electronic design application of a **sc_mutex** might be arbitration for a shared bus. Here the ability of multiple masters to access the bus must be controlled. In lieu of an arbiter design, the **sc_mutex** might be used to manage the bus resource quickly until an arbiter can be arranged or designed.

In fact, the mutex might even be part of the class implementing the bus model as illustrated in the following example:

```
class bus {
  sc_mutex bus_access;
  ...
  void write(int addr, int data) {
    bus_access.lock();
    // perform write
    bus_access.unlock();
  }
  ...
};//endclass
```

*Figure 8-4.* Example of sc_mutex Used in Bus Class

Used with an **SC_METHOD** process, access might look like this:

```
void grab_bus_method() {
  if (bus_access.trylock()) {
    /* access bus */
  }//endif
}
```

*Figure 8-5.* Example of sc_mutex with an sc_method

One downside to the **sc_mutex** is the lack of an event that signals when an **sc_mutex** is freed, which requires using **trylock** repeatedly based on some other event or time based delay. Remember, unless your process waits (via a **wait** or **return**), you will not allow the process that currently owns the resource to free the resource, and the simulation will hang.

## 8.3  sc_semaphore

For some resources, you can have more than one copy or owner. A good example of this would be parking spaces in a parking lot.

To manage this type of resource, SystemC provides the **sc_semaphore**. When creating an **sc_semphore** object, it is necessary to specify how many are available. In a sense, a mutex is merely a semaphore with a count of one.

An **sc_semaphore** access consists of waiting for an available resource and then posting notice when finished with the resource.

```
sc_semaphore NAME(COUNT);

//To lock a mutex, NAME (wait until
// unlocked if in use)
NAME.wait();

// Non-blocking, returns true if success else false
NAME.trywait()

//Returns number of available semaphores
NAME.get_value()

//To free a previously locked mutex
NAME.post();
```

*Figure 8-6.* Syntax of sc_semaphore

It is important to realize that the **sc_semaphore::wait**() is a distinctly different method from the **wait**() method previously discussed in conjunction with **SC_THREAD**s. In fact, under the hood, the **sc_semaphore::wait**() is implemented with the **wait**(event).

A modern gas station with self-service would be a good example for using semaphores. A semaphore could represent the number of available gas pumps.

```
SC_MODULE(gas_station) {
  sc_semaphore pump(12);
  void customer1_thread {
    for(;;) {
      // wait till tank empty
      ...
      // find an available gas pump
      pump.wait();
      // fill tank & pay
  }
};
```

*Figure 8-7.* Example of sc_semaphore—gas_station

A multi-port memory model is a good example of an electronic system-level design application for **sc_semaphore**. You might use the semaphore to indicate the number of read or write accesses allowed.

```
class multiport_RAM {
  sc_semaphore read_ports(3);
  sc_semaphore write_ports(2);
  ...
  void read(int addr, int& data) {
    read_ports.wait();
    // perform read
    read_ports.post();
  }
  void write(int addr, int data) {
    write_ports.lock();
    // perform write
    write_ports.unlock();
  }
  ...
};//endclass
```

*Figure 8-8.* Example of sc_semaphore—multiport_RAM

Other examples might include allocation of timeslots in a TDM (time division multiplex) scheme used in telephony, controlling tokens in a token ring, or perhaps even switching information to obtain better power management.

## 8.4  sc_fifo

Probably the most popular channel for modeling at the architectural level is the **sc_fifo**. First-in first-out queues (i.e., FIFOs) are a common data structure used to manage data flow. FIFOs are some of the simplest structures to manage.

In the very early stages of architectural design, the unbounded[25] STL **deque<>** (double ended queue) provides an easy implementation of a FIFO. Later, when bounds are determined or reasonable guesses at FIFO depths and SystemC elements come into stronger play, the **sc_fifo<>** may be used.

---

[25] Limited only by the resources of the simulation machine itself.

By default, an **sc_fifo<>** has a depth of 16. The data type (i.e., **typename**) of the elements also needs to be specified. An **sc_fifo** may contain any data type including large and complex structures (e.g., a TCP/IP packet or a disk block).

```
sc_fifo<ELEMENT_TYPENAME> NAME(SIZE);

NAME.write (VALUE);
NAME.read (REFERENCE);
… = NAME.read () /* function style */
if (NAME.nb_read (REFERENCE)) { // Non-blocking
                              // true if success

  …
}
if (NAME.num_available() == 0)
  wait(NAME.data_written_event());
if (NAME.num_free() == 0)
  next_trigger(NAME.data_read_event());
```

*Figure 8-9.* Syntax of sc_fifo—Abbreviated

For example, FIFOs may be used to buffer data between an image processor and a bus, or a communications system might use FIFOs to buffer information packets as they traverse a network.

Some architectural models are based on Kahn process networks[26] for which unbounded FIFOs provide the interconnect fabric. Given an appropriate depth, you can use **sc_fifo** for this purpose as illustrated in the next very simple example.

---

[26] Kahn, G. 1974. The semantics of a simple language for parallel programming, in Proc. IFIP74, J.L. Rosenfeld (ed.), North-Holland, pp.471-475.

```
SC_MODULE(kahn_ex) {
  ...
  sc_fifo<double> a, b, y;
  ...
};
// Constructor
kahn_ex::kahn_ex() : a(10), b(10), y(20)
{
  ...
}
void kahn_ex::addsub_thread() {
  for(;;) {
    y.write(kA*a.read() + kB*b.read());
    y.write(kA*a.read() - kB*b.read());
  }//endforever
}
```

*Figure 8-10.* Example of sc_fifo kahn_ex

Software uses for FIFOs are numerous and include such concepts as mailboxes and other types of queues.

Note that when considering efficiency, passing pointers to large objects is most efficient. Be sure to consider using a safe pointer object if using a pointer. The **shared_ptr**<> of the GNU publicly licensed Boost library, http://www.boost.org, makes implementation of smart pointers very straight forward.

Generally speaking, the STL may be more suited to software FIFOs. The use of the STL **deque**<> might be used to manage an unknown number of stimulus data from a test bench.

In theory, an **sc_fifo** could be synthesized at a behavioral level. It currently remains for a synthesis tool vendor to provide the functionality.

## 8.5 Exercises

For the following exercises, use the samples provided in www.EklecticAlly.com/Book/.

**Exercise 8.1:** Examine, predict the output, compile, and run `mutex_ex`.

**Examine 8.2:** Examine, compile, and run `semaphore_ex`. Add another family member. Explain discrepancies in behavior.

**Exercise 8.3:** Examine, compile, and run `fifo_fir`. Add a second filter stage to the network.

**Exercise 8.4:** Examine, compile, and run `fifo_of_ptr`. Discuss how one might compensate for the simulated transfer of a large packet.

**Exercise 8.5:** Examine, compile, and run `fifo_of_smart_ptr`. Notice the absence of **delete**.

| Predefined Primitive Channels: Mutexes, FIFOs, & Signals | | |
|---|---|---|
| Simulation Kernel | Threads & Methods | Channels & Interfaces | Data types: Logic, |
| | Events, Sensitivity & Notifications | Modules & Hierarchy | Integers, Fixed point |

Chapter 9

# EVALUATE-UPDATE CHANNELS
## *SC_SIGNALS*

The preceding chapter considered synchronization mechanisms common to software. This chapter delves into electronic hardware[27].

Electronic signals behave in a manner approaching instantaneous activity. Generally, electronic signals have a single source (producer), but multiple sinks (consumer). It is quite important that all sinks "see" a signal update at the same time.

The easiest way to understand this concept is to consider the common hardware shift register. This model has a number of registers or memory elements as indicated in the following diagram.

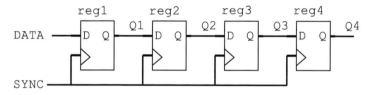

*Figure 9-1.* Shift Register

[27]. It is unclear whether the concepts discussed here have any application outside of electrical signals.

Data moves from left to right synchronous to the clock labeled SYNC. In software (e.g., C/C++), this would be modeled with four ordinary assignments:

```
Q4 = Q3;
Q3 = Q2;
Q2 = Q1;
Q1 = DATA;
```

*Figure 9-2.* Example of Modeling a Software-Only Shift Register

For this register to work, ordering is very important. In hardware, things are more difficult. Each register (reg1...reg4) is an independent concurrent process. Recall that the simulator places no order requirements on the processes. Below is a diagram from Chapter 7, Concurrency. Consider each process to represent a register from the preceding design.

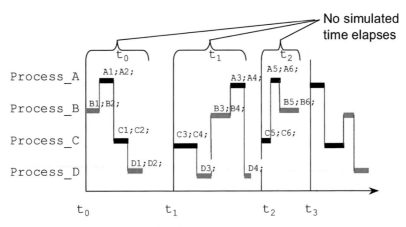

*Figure 9-3.* Simulated Activity with Four Concurrent Processes

Since there is no guarantee that one assignment will take place before the other, we need to find some other solution. One idea might be to use events to force an ordering. This design would have Process_A wait for an event from Process_B before assigning its register, Process_B wait for an event from Process_C before assigning its register, and so on. This design requires coding both a **wait** and **notify** for each register, and it can become quite tiresome.

Another solution involves representing each register as a two-deep FIFO. This approach seems unnecessarily complex requiring two storage locations for the data and two pointers/counters to manage the state of the FIFO.

## 9.1 Completed Simulation Engine

To solve this problem, simulators have a feature known as the evaluate-update paradigm. The following diagram is the complete SystemC simulation kernel with the added update phase. It is possible to go from evaluate to update and back. This cycle is known as the delta-cycle. Even when the simulator moves from evaluate to update to advance time, we say that at least one delta-cycle has occurred. Let's see how the delta-cycle is used.

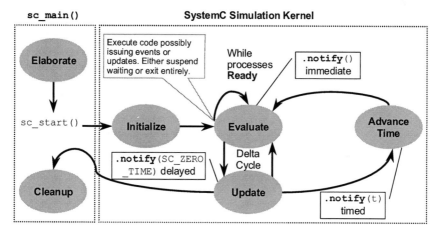

*Figure 9-4.* Full SystemC Simulation Engine

Special channels, known as signal channels use this update phase as a point of data synchronization. To accomplish this synchronization, every channel has two storage locations: the current value and the new value. Visually, there are two sets of data: new and current.

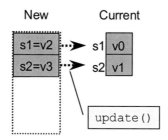

*Figure 9-5.* Signal Channel Data Storage

When a process writes to a signal channel, the process stores into the new value rather than into the current value, and the process calls

**request_update**() to cause the simulation kernel to call the channel's **update**() method during the update phase.

After the evaluate phase has completed (i.e., there are no more processes in the ready state), the kernel calls the **update**() method for each channel instance that requested an update. The **update**() method may do more than simply copy the new value into the current value. It may resolve contentions and notify **sc_event**s (e.g., to indicate a change and wake up a process in the waiting state).

An important aspect of this paradigm is the current value remains unchanged. If a process writes to an evaluate-update channel and then immediately (i.e., without suspending) accesses the channel, it will find the current value unchanged.

This behavior is a frequent cause for confusion for simulation neophytes. Those familiar with HDLs should not be surprised. Shortly, we will discuss techniques to make this less of a surprise.

Another consideration is that the new value will contain only the last value written during a given delta cycle. Thus, writing repeatedly overwrites the previous new value.

## 9.2 sc_signal, sc_buffer

The **sc_signal**<> primitive channel and its close relative, **sc_buffer**<>, both use the evaluate-update paradigm. Here is the syntax for declaration, reading, and writing:

```
sc_signal<datatype> signame[, signameᵢ]…;
signame.write(newvalue);
signame.read(varname);
sensitive << signame.default_event();
wait(signame.default_event()|...);
if (signame.event() == true) {
  // occurred in previous delta-cycle
```

*Figure 9-6.* Syntax of sc_signal

The **write()** method contains the evaluate phase portion of the evaluate-update behavior. The **write()** method includes a call to the protected **sc_prim_channel::request_update()** method. The call to **sc_signal::update()** is hidden, and the call occurs during the update phase when the kernel calls it as a result of the **request_update**.

The **event()** method is special. Normally, it is impossible to determine which event caused a return from **wait()**; however, for **sc_signal** channels (including other derivatives mentioned in this chapter), the **event()** method may be called to see if the channel issued an event in the immediately previous delta-cycle. This ability to determine which event caused the last return from **wait()** does not preclude the occurrence of other events in the previous cycle.

It should be noted that **sc_signal**<> is essentially identical to VHDL's **signal**. For Verilog, the analogy is a **reg** that uses non-blocking assignments (**<=**) exclusively.

An example of usage and the slightly surprising results[28] are in order.

---

[28] This usage may surprise non-HDL experienced folks. HDL-experienced users should understand the VHDL or Verilog analogy.

```
// Declare variables
int               count;
sc_string         message_temp;
sc_signal<int>    count_sig;
sc_signal<sc_string> message_sig;

// Initializing during 1st delta cycle
count_sig.write(10);
message_sig.write("Hello");
count = 11;
message_temp = "Whoa";
cout << "count is " << count << " "
     << "count_sig is " << count_sig << endl
     << "message_temp is '" << message_temp << "' "
     << "message_sig is '" << message_sig << "'"
     <<endl
     << "Waiting" << endl;
wait(SC_ZERO_TIME);

// 2nd delta cycle
count = 20;
count_sig.write(count);
cout << "count is " << count << " "
     << "count_sig is " << count_sig << endl
     << "message_temp is '" << message_temp << "' "
     << "message_sig is '" << message_sig << "'"
     <<endl
     << "Waiting" << endl;
wait(SC_ZERO_TIME);

// 3rd delta cycle
message_sig.write(message_temp = "Rev engines");
cout << "count is " << count << " "
     << "count_sig is " << count_sig << endl
     << "message_temp is '" << message_temp << "' "
     << "message_sig is '" << message_sig << "'"
     <<endl
     << "Done" << endl;
```

*Figure 9-7.* Example of sc_signal

The example in *Figure 9-7* produces the result shown in *Figure 9-8*. Notice how the current value remains unchanged until a delta-cycle has occurred.

```
count is 11 count_sig is 0
message_temp is 'Whoa' message_sig is ''
Waiting
count is 20 count_sig is 10
message_temp is 'Whoa' message_sig is 'Hello'
Waiting
count is 20 count_sig is 20
message_temp is 'Rev engines' message_sig is 'Hello'
Waiting
```

*Figure 9-8.* Example of sc_signal Output

Because the code uses a naming convention (i.e., appended _sig to the signals), it is relatively easy to spot the evaluate-update signals and make the mental connection to the behavior. Without the naming convention, one might wonder if the identifiers represent some other channel (e.g., **sc_fifo<>**).

In addition to the preceding syntax, SystemC has overloaded the assignment and copy operators to allow the following *dangerous* syntaxes:

```
varname = signame.read();
signame = newvalue;
varname = signame;
```

*Figure 9-9.* Syntax of sc_signal (Dangerous)

The reason we consider these syntaxes dangerous relates to the issue of the evaluate-update paradigm. Consider the following example:

```
// Convert rectangular to polar coordinates
r = x;
if ( r != 0 && r != 1 ) r = r * r;
if ( y != 0 ) r = r + y*y;
cout << "Radius is " << sqrt(r) << endl;
```

*Figure 9-10.* Dangerous sc_signal

Without sufficient context, the casual reader would be quite surprised at the results shown below. Assume on entry x=3, y=4, r=0.

```
Radius is 0
```

*Figure 9-11.* Example of sc_signal Output (Dangerous)

Even when using what might be considered the safer syntax, you must be careful. We strongly suggest that you use a naming style.

One beneficial aspect of **sc_signal<>** and **sc_buffer<>** channels is a restriction that only a single process may write to a given signal during a specific delta-cycle. This restriction avoids the potential danger of two processes non-deterministically asserting a value and creating a race condition. A runtime error is flagged in this situation if and only if you have defined the compile-time macro **DEBUG_SYSTEMC**. You can find an example of this danger in the danger_ex example.

## 9.3   sc_signal_resolved, sc_signal_rv

There are times when it is appropriate to have multiple writers. One of these situations involves modeling buses that have the possibility of high impedance (i.e., Z) and contention (i.e., X).

**Multiple Drivers on a Bus**

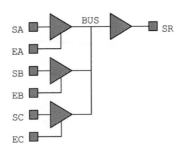

*Figure 9-12.* Tri-State Bus

SystemC provided the specialized channels **sc_signal_resolved** and **sc_signal_rv<>**[29].

---

[29]  **_rv** means resolved vector.

```
sc_signal_resolved name;
sc_signal_rv<WIDTH> name;
```

*Figure 9-13.* Syntax of sc_signal_resolved and sc_signal_rv

The base functionality has identical semantics to `sc_signal<sc_logic>`; however, it allows for multiple writers and provides built-in resolution functionality as follows:

*Table 9-1* Resolution Functionality for sc_signal_resolved

| A\B | '0' | '1' | 'X' | 'Z' |
|-----|-----|-----|-----|-----|
| '0' | '0' | 'X' | 'X' | '0' |
| '1' | 'X' | '1' | 'X' | '1' |
| 'X' | 'X' | 'X' | 'X' | 'X' |
| 'Z' | '0' | '1' | 'X' | 'Z' |

One minor failing of SystemC is the lack of direct support for several common system-level bus concepts. Specifically, SystemC has no mechanisms for pull-ups, pull-downs, nor various open-source or open-drain variations.

For these, you have to create your own channels, which is not too difficult. The easiest way is to create a class derived from an existing class that almost works. Here is the resolution table for a pull-up functionality.

*Table 9-2.* Resolution Functionality for sc_signal_resolved

| A\B | '0' | '1' | 'X' | 'Z' |
|-----|-----|-----|-----|-----|
| '0' | '0' | 'X' | 'X' | '0' |
| '1' | 'X' | '1' | 'X' | '1' |
| 'X' | 'X' | 'X' | 'X' | 'X' |
| 'Z' | '0' | '1' | 'X' | '1' |

Notice that there is only one difference in the table (shaded). The custom channel in *Figure 9-14* implements this resolution for a single-bit pull-up functionality.

```
struct ea_signal_pullup: public sc_signal_resolved {
 ea_signal_pullup() {}
 explicit ea_signal_pullup(const char* nm )
   :sc_signal_resolved(nm) {}
 virtual void update() {
   sc_logic_resolve::resolve( m_new_val, m_val_vec );
   if (m_new_val == SC_LOGIC_Z) {
     m_new_val = SC_LOGIC_1;
   }
   base_type::update();
 }// end update
};//endstruct ea_signal_pullup
```

*Figure 9-14.* Example of ea_signal_pullup

Note that the protected method **sc_logic_resolve::resolve**() is not currently documented in the LRM and could theoretically be removed. We hope it will be added to the final specification considering how prevalent concepts such as pull-ups are at the system level. For more information, study the source code of the Open SystemC Initiative reference implementation.

## 9.4   Template Specializations of sc_signal Channels

SystemC has several template specializations that bear discussion. A template specialization occurs when a definition is provided for a specific template value. If there is more than one template variable involved, we call it a partial specialization.

For example, **sc_signal** has a single template variable representing the **typename**. SystemC defines some additional behaviors for **sc_signal<bool>** that are not available for the general case. Thus, an **sc_signal<char>** does not support the concept of a **posedge_event**().

The specialized templates **sc_signal<bool>** and **sc_signal<sc_logic>** have the following extensions:

```
sensitive << signame.posedge_event()
         << signame.negedge_event();
wait(signame.posedge_event()
    |signame.negedge_event());
if (signame.posedge_event()
    |signame.negedge_event())  {
```

*Figure 9-15.* Syntax of Specializations posedge and negedge

For **sc_logic**, a **posedge_event** occurs on any transition to **SC_LOGIC_1**, which includes **SC_LOGIC_X** and **SC_LOGIC_Z**. The same is true of transitions to **SC_LOGIC_0** and the **negedge_event**.

The Boolean **posedge()** and **negedge()** methods apply similarly to the **event()** method, and they only apply to the immediately previous delta-cycle.

It is notable that **sc_buffer** does *not* support these specializations.

## 9.5  Exercises

For the following exercises, use the samples provided in www.EklecticAlly.com/Book/.

**Exercise 9.1:** Examine, compile, and run `signal_ex`.

**Exercise 9.2:** Examine, compile, and run `buffer_ex`.

**Exercise 9.3:** Examine, compile, and run `danger_ex`.

**Exercise 9.4:** Examine, compile, and run `resolved_ex`. Observe the definition of **DEBUG_SYSTEMC** in `../Makefile.defs`.

**Exercise 9.5:** Examine the interactive simulation illustrating the simulation engine under the web page www.EklecticAlly.com under Library Tutorials SimulationEngine.

| | | Predefined Primitive Channels: Mutexes, FIFOs, & Signals | | |
| --- | --- | --- | --- | --- |
| | Simulation Kernel | Threads & Methods | Channels & Interfaces | Data types: Logic, Integers, Fixed point |
| Chapter 10 | | Events, Sensitivity & Notifications | **Modules & Hierarchy** | |

# STRUCTURE
*Design Hierarchy*

This chapter describes SystemC's facilities for implementing structure, sometimes known as design hierarchy. Design hierarchy concerns both the hierarchical relationships of modules discussed here and the connectivity that allows modules to communicate in an orderly fashion. Connectivity will be discussed in Chapter 11.

## 10.1 Module Hierarchy

Thus far, we have only examined modules containing a single level of hierarchy with all processes residing in a single module. This level of complexity might be acceptable for small designs, but real system designs require partitioning and hierarchy for understanding and project management. Project management includes documentation and practical issues such as integration of third-party intellectual property.

Design hierarchy in SystemC uses instantiations of modules as member data of parent modules. In other words, to create a level of hierarchy, create an **sc_module** object that represents the sub-module within the parent module.

Consider the hierarchy in *Figure 10-1* for the following discussions and examples. In this case, we have a parent module named **Car**, and a sub-module named **Engine**. To obtain the hierarchical relationship, we create an **Engine** object and a **body** object within the definition of the **Car** class.

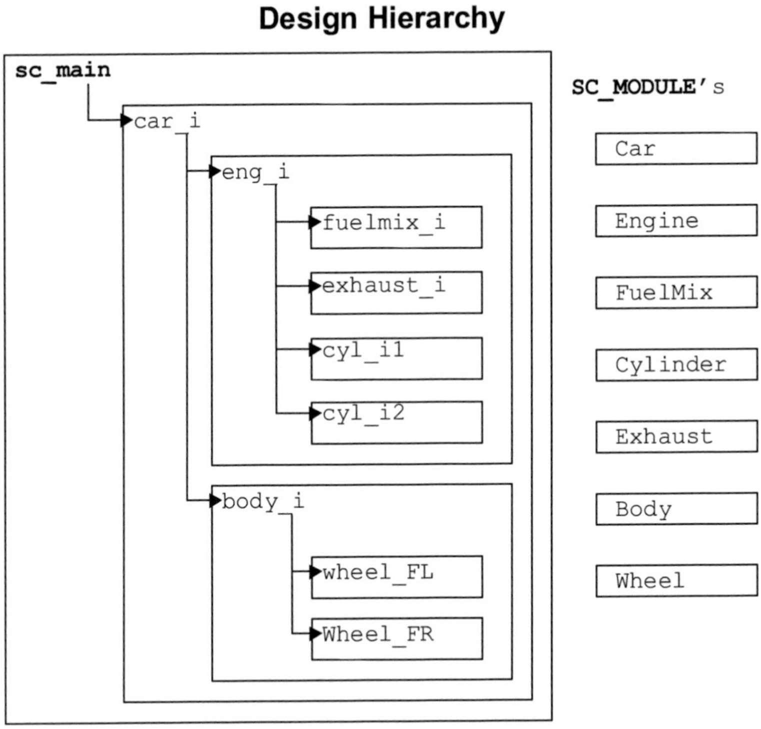

## Design Hierarchy

*Figure 10-1.* Module Hierarchy

Module instantiation occurs inside the constructor. You can define the code that implements the constructor in either the header file or the implementation file. Since hierarchy reflects an internal implementation decision, the authors prefer to see the constructor defined in the implementation. The only time this cannot be done this way is when defining a templated module, a relatively rare occurrence, due to compiler restrictions[30].

C++ offers two basic ways to instantiate modules. First, a module object may be created directly by declaration. Alternatively, a module object may be indirectly referenced by means of a pointer and dynamic allocation.

Creation of hierarchy at the top level (**sc_main**) is slightly different from instantiation within modules. This difference results from differences in C++ syntax requirements for initialization outside of a class definition.

---

[30] A future version of C++ compiler/linker toolsets may fix this restriction.

Because it is likely you may see any combination of these approaches, we will illustrate all six approaches:

- Direct top-level

- Indirect top-level

- Direct sub-module header-only

- Direct sub-module

- Indirect sub-module header-only

- Indirect sub-module

There are likely a few more variants, but understanding these should suffice.

## 10.2 Direct Top-Level Implementation

First, we illustrate top-level with direct instantiation, which has already been presented in Hello_SystemC and is used in all the succeeding discussions. It is simple and to the point. Sub-design instances are simply instantiated and initialized in one statement.

```
//FILE: main.cpp
  #include "Wheel.h"
  int sc_main(int argc, char* argv[]) {
    Wheel wheel_FL("wheel_FL");
    Wheel wheel_FR("wheel_FR");
    sc_start();
  }
```

*Figure 10-2.* Example of main with Direct Instantiation

## 10.3 Indirect Top-Level Implementation

A minor variation on this approach, main with indirect instantiation, places the design on the heap and adds two lines of syntax with both a pointer declaration and an instance creation via **new**. This variation takes more code; however, it adds the possibility of dynamically configuring the design with the addition of if-else and looping constructs.

A design with a regular structure might even construct an array of designs and programmatically instantiate and connect them up. We will discuss connectivity in the next section of this chapter.

```
//FILE: main.cpp
  #include "Wheel.h"
  int sc_main(int argc, char* argv[]) {
    Wheel* wheel_FL; // pointer to FL wheel
    Wheel* wheel_FR; // pointer to FR wheel
    wheel_FL = new Wheel("wheel_FL"); // create FL
    wheel_FR = new Wheel("wheel_FR"); // create FR
    sc_start();
    delete wheel_FL;
    delete wheel_FR;
  }
```

*Figure 10-3.* Example of main with Indirect Instantiation

## 10.4 Direct Sub-Module Header-Only Implementation

When dealing with sub-modules (i.e., beneath or within a module), things become mildly more interesting because C++ semantics require use of an initializer list for the direct approach:

```
//FILE:Body.h
  #include "Wheel.h"
  SC_MODULE(Body) {
    Wheel wheel_FL;
    Wheel wheel_FR;
    SC_CTOR(Body)
    : wheel_FL("wheel_FL"), //initialization
      wheel_FR("wheel_FR")  //initialization
    {
      // other initialization
    }
  };
```

*Figure 10-4.* Example of Direct Instantiation in Header

## 10.5 Indirect Sub-Module Header-Only Implementation

Use of indirection renders the instantiation a little bit easier to read for the sub-module header-only case; however, no other advantages are clear.

```
//FILE:Body.h
  #include "Wheel.h"
  SC_MODULE(Body) {
    Wheel* wheel_FL;
    Wheel* wheel_FR;
    SC_CTOR(Body) {
      wheel_FL = new Wheel("wheel_FL");
      wheel_FR = new Wheel("wheel_FR");
      // other initialization
    }
  };
```

*Figure 10-5.* Example of Indirect Instantiation in Header

## 10.6 Direct Sub-Module Implementation

One disadvantage of the preceding approach is that it exposes the complexities of the constructor body to all potential users. Moving the constructor into the implementation (i.e., the *module*.cpp file) requires the use of **SC_HAS_PROCESS**.

```
//FILE:Body.h
  #include "Wheel.h"
  SC_MODULE(Body) {
    Wheel wheel_FL;
    Wheel wheel_FR;
    SC_HAS_PROCESS(Body);
    Body(sc_module_name nm);
  };
```

```
//FILE: Body.cpp
  #include "Body.h"
  // Constructor
  Body::Body(sc_module_name nm)
  : wheel_FL("wheel_FL"),
    wheel_FR("wheel_FR"),
    sc_module(nm)
  {
    // other initialization
  }
```

*Figure 10-6.* Example of Direct Instantiation and Separate Compilation

## 10.7 Indirect Sub-Module Implementation

Moving the module indirect approach into the implementation file has the advantage of possibly supplying pre-compiled object files, and this approach may be good for intellectual property (IP) distribution. This advantage is in addition to the possibility of dynamically determining the configuration discussed previously.

```
//FILE:Body.h
  struct Wheel;
  SC_MODULE(Body) {
    Wheel* wheel_FL;
    Wheel* wheel_FR;
    SC_HAS_PROCESS(Body);
    Body(sc_module_name nm); // Constructor
  };
```

```
//FILE: Body.cpp
  #include "Wheel.h"
  // Constructor
  Body::Body(sc_module_name nm)
  : sc_module(nm)
  {
    wheel_FL = new Wheel("wheel_FL");
    wheel_FR = new Wheel("wheel_FR");
    // other initialization
  }
```

*Figure 10-7.* Example of Indirect Separate Compilation

Notice the absence of #include Wheel.h in Body.h of the preceding example. This omission could be a real advantage when providing Body for use by another group (such as IP). You need to provide only Body.h and a compiled object or library (e.g., Body.o or Body.a) files. You can then develop your implementation independently. This approach is good for both internal and external IP distribution.

## 10.8 Contrasting Implementation Approaches

The following table contrasts the features of the six approaches.

*Table 10-1.* Comparison of Hierarchical Instantiation Approaches

| Level | Allocation | Pros | Cons |
|---|---|---|---|
| Main | Direct | Least code | Inconsistent with other levels |
| Main | Indirect | Dynamically configurable | Involves pointers |
| Module | Direct header only | All in one file  Easier to understand | Requires sub-module headers |
| Module | Indirect header only | All in one file  Dynamically configurable | Involves pointers  Requires sub-module headers |
| Module | Direct with separate compilation | Hides implementation | Requires sub-module headers |
| Module | Indirect with separate compilation | Hides sub-module headers and implementation  Dynamically configurable | Involves pointers |

Some groups have the opinion that the top-level module should instantiate a single design with a fixed name (e.g., Design_top) and then only deal with the lower levels in a consistent fashion. Some EDA tools perform all this magic for you.

## 10.9 Exercises

For the following exercises, use the samples provided at www.EklecticAlly.com/Book/.

**Exercise 10.1:** Examine, compile, and run the sedan example. Which styles are simplest?

**Exercise 10.2:** Examine, compile, and run the convertible example. Notice the forward declarations of Body and Engine. How might this be an advantage when providing IP?

Chapter 11

# COMMUNICATION
*Ports*

This chapter describes SystemC's facilities for implementing connectivity, which enables orderly communication between modules.

## 11.1 Communication: The Need for Ports

Hierarchy without the ability to communicate between modules is not very useful, but what is the best way to communicate? There are two concerns: safety and ease of use. Safety is a concern because all activity occurs within processes, and care must be taken when communicating between processes to avoid race conditions. Events and channels are used to handle this concern.

Ease of use is more difficult to address. Let us dispense with any solution involving global variables, which are well known as a poor methodology. Another possibility is to have a process in an upper-level module. This process would monitor and manage events defined in instantiated modules. This mechanism is awkward at best.

SystemC takes an approach that lets modules use channels inserted between the communicating modules. SystemC accomplishes this communication with a concept called a port. Basically, a port is a pointer to a channel outside the module.

Consider the following example:

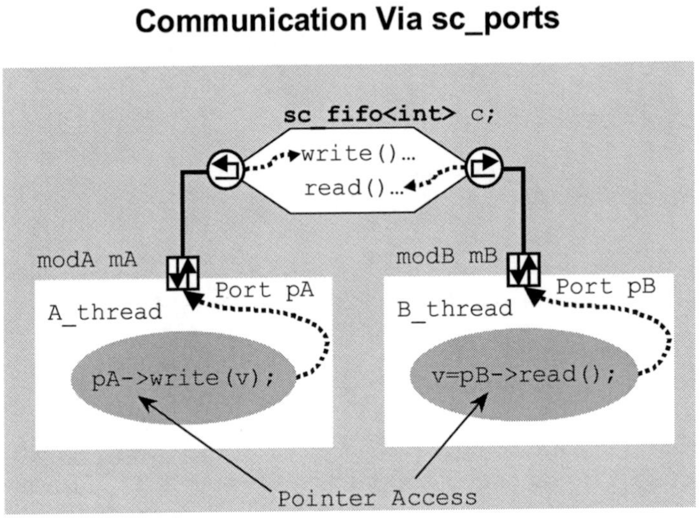

## Communication Via sc_ports

*Figure 11-1.* Communication Via Ports

The process A_thread in module modA communicates a value contained in local variable **v** by calling the **write** method of the parent module's channel **c**. Process B_thread in module modB may retrieve a value via the **read** method of channel c.

Notice that access is accomplished via pointers pA and pB. Notice also that the two processes really only need access to the methods **write** and **read**. More specifically, modA only needs access to **write**, and modB only needs access to **read**. This separation of access is managed using a concept known as an interface, which is described in the next section.

## 11.2 Interfaces: C++ and SystemC

C++ defines a concept known as an abstract class. An abstract class is a class that is never used directly, but it is used only via derived sub-classes. Partly to enforce this concept, abstract classes usually contain pure virtual functions. Pure virtual functions are not allowed to provide an implementation in the abstract class where they are defined as pure. This restriction in turn compels any class derived from the abstract class to override all the pure virtual functions, or in other words, the class derived from the abstract class must provide an implementation for all the pure virtual functions.

The following diagram illustrates the concept. Pure virtual functions are identified by 1) the keyword **virtual** and 2) the **=0;** to indicate they're pure.

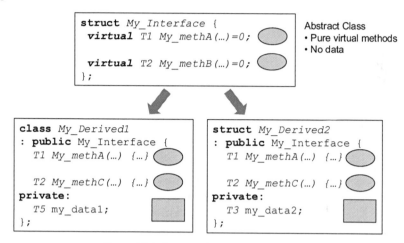

*Figure 11-2.* C++ Interface Class Relationships

If a class contains no data members and only contains pure virtual methods, it is known as an interface class. Here is a short example of an interface class:

```
struct my_interface {
    virtual void write(unsigned addr, int data) = 0;
    virtual int read(unsigned addr) = 0;
};
```

*Figure 11-3.* Example of C++ Interface

The concept of interfaces has a powerful property when used with polymorphism. Recall from C++ that polymorphism is the following idea: A derived class can be processed by a member function of the parent class. For example, if a member function simply determines the number of "elements" in a module, it doesn't matter what the elements are in the module.

Consider the preceding figure of C++ interface class relationships. A function using My_Interface might access My_methA(). If the current object is of class My_Derived2, then the actual My_methA() call results in My_Derived2::My_methA().

If an object is declared as a pointer to an interface class, it may be used with multiple derived classes. Suppose we define two derived classes as follows:

```
struct multiport_memory_arch: public my_interface {
  virtual void write(unsigned addr, int data) {
    mem[addr] = data;
  }// end write
  virtual int read(unsigned addr) ) {
    return mem[addr];
  }//end read
private:
  int mem[1024];
};
```

```
struct multiport_memory_RTL: public my_interface {
  virtual void write(unsigned addr, int data) {
    // complex details of RTL memory write
  }// end write
  virtual int read(unsigned addr) ) {
    // complex details of RTL memory read
  }// end read
private:
  // complex details of RTL memory storage
};
```

*Figure 11-4.* Example of Two Derivations From Interface Class

Suppose now we write some code to access the classes derived above.

```
void memtest(my_interface mem) {
  // complex memory test
}

multiport_memory_arch fast;
multiport_memory_RTL  slow;
memtest(fast);
memtest(slow);
```

*Figure 11-5.* Example of C++ Interface

As seen in the preceding example, the same code may access multiple variations of a design. You can think of an interface as the application programming interface (API) to a set of derived classes. This same concept is used in SystemC to implement ports.

DEFINITION: A SystemC interface is an abstract class that inherits from **sc_interface** and provides only pure virtual declarations of methods referenced by SystemC channels and ports. No implementations or data are provided in a SystemC interface.

We now provide the concise definition of the SystemC channel.

DEFINITION: A SystemC channel is a class that implements one or more SystemC interface classes and inherits from either **sc_channel** or **sc_prim_channel**. A channel implements all the methods of the inherited interface classes.

By using interfaces to connect channels, we can implement modules independent of the implementation details of the communication channels.

Consider the following diagram:

## Power of Interfaces

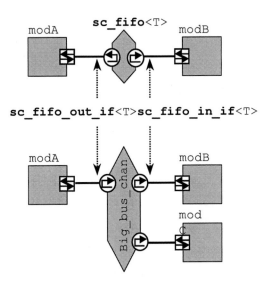

*Figure 11-6.* The Power of Interfaces

In one design, modules **modA** and **modB** are connected via a FIFO. With no change to the definition of **modA** or **modB**, we can swap out the FIFO for a different channel. All that matters is for the interfaces used to remain constant. In this example, the interfaces are `sc_fifo_out_if<T>` and `sc_fifo_in_if<T>`. In the next few sections, the mechanics of using interfaces are described.

## 11.3 Simple SystemC Port Declarations

Given the definition of an interface, we now present the definition of a port.

DEFINITION   A SystemC port is a class templated with and inheriting from a SystemC interface. Ports allow access of channels across module boundaries.

Specifically, the syntax of a simple SystemC port follows:

```
sc_port<interface> portname;
```

*Figure 11-7.* Syntax of Basic sc_port

SystemC ports are always defined within the module class definition. Here is a simple example:

```
SC_MODULE(stereo_amp) {
  sc_port<sc_fifo_in_if<int> >  soundin_p;
  sc_port<sc_fifo_out_if<int> > soundout_p;

  ...
};
```

*Figure 11-8.* Example of Defining Ports within Module Class Definition

Notice the extra blank space following the greater-than symbol (>). This is required C++ syntax when nesting templated classes.

## 11.4 Many Ways to Connect

Given the declaration of a port, we now address the issue of connecting ports, channels, modules, and processes. The following diagram illustrates the types of connections that are possible with SystemC:

## Port Connections

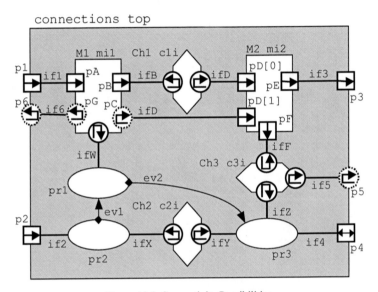

*Figure 11-9.* Connectivity Possibilities

This diagram is quite busy. Let's examine the pieces by name, and then discuss the rules of interconnection.

First, there are three modules represented with rectangles. The enclosing module instance is named `top`. The two sub-module instances within `top` are named `mi1` and `mi2`.

Each of the modules has one or more ports represented with squares. Directional arrows within the ports indicate the primary flow of information. The ports for `top` are `p1`, `p2`, `p3`, `p4`, `p5`, and `p6`, which use interfaces named `if1`, `if2`, `if3`, `if4`, `if5`, and `if6`, respectively.

The ports for `mi1` are `pA`, `pB`, `pC`, and `pG`, which are connected to interfaces named `if1`, `ifB`, `ifD`, and `if6`, respectively.

Module `M1` also provides interfaces `ifW` and `if6`.

The ports for mi2 are pD[0], pD[1], pE, and pF, which are connected to interfaces named if3, ifD, and ifF, respectively.

Next, three instances of channels represented with hexagonal shapes exist within top. These are named c1i, c2i, and c3i.

Each channel implements one or more interfaces represented by circles with a bent arrow. The arrow is intended to indicate the possibility of a call returning a value. It is possible for a channel to implement only a single interface. Channel c1i implements interfaces ifB and ifD. Channel c2i implements interfaces ifX and ifY. Finally, channel c3i implements interfaces if5, ifF, and ifZ.

Last, there are three processes named pr1, pr2, and pr3. For this description, we don't need to know what type of processes (i.e., threads vs methods). There are two explicit events, ev1 and ev2 used for signaling between processes.

From this diagram, several rules may be observed. As we already know, processes may communicate with other processes at the same level either via channels or synchronized via events. Processes may communicate with processes outside the local design module through ports bound to channels by way of interfaces. Processes may also communicate with processes in sub-module instances via interfaces to channels connected to the sub-module ports or by way of interfaces through the module itself of an **sc_export** (to be discussed later). Any other attempt at inter-process communication is either forbidden or dangerous.

Ports may connect via interfaces only to local channels, ports of sub-modules, or to processes indirectly.

There are a few interesting features that will be discussed later. First, module instance mi1 implements an interface ifW. We will discuss this topic in Chapter 13, in Custom Channels.

Second, port pD appears to be an array of size 2. We will discuss port arrays in Chapter 12, More on Ports.

Third, port p5 and port pC illustrate a new type of port for SystemC version 2.1 known as an **sc_export**. We will discuss this topic in the next page chapter.

As a summary, let's view this information in a tabular format.

*Table 11-1* Ways to Interconnect

| From | To | Method |
|------|-----|--------|
| Port | Sub-module | Direct connect via `sc_port` |
| Process | Port | Direct access by process |
| Sub-module | Sub-module | Local channel connection |
| Process | Sub-module | Local channel connection –or- via `sc_export` –or- interface implemented by sub-module[31] |
| Process | Process | Events or local channel |
| Port | Local channel | Direct connect via `sc_export` |

## 11.5 Port Connection Mechanics

Modules are connected to channels after both the modules and channels have been instantiated. There are two syntaxes for connecting ports: by-name and by-position. Due to the error-prone nature of positional notation (especially since the number of ports on modules tends to be large and changes), the authors strongly prefer connection by-name. Here are both syntaxes.

```
mod_inst.portname(channel_instance); // Named
mod_instance(channel_instance,…); // Positional
```

*Figure 11-10.* Syntax of Port Connectivity

An example should help greatly. We'll use a simple video mixer example with a color space transformation. For this example, we will use two standard SystemC interface classes, **sc_fifo_in_if** and **sc_fifo_out_if**, which support **read**() and **write**(value), respectively. First, we introduce the module definitions.

---

[31] In this case, the module is also known as a "hierarchical channel", which will be discussed later.

```
//FILE: Rgb2YCrCb.h
SC_MODULE(Rgb2YCrCb) {
  sc_port<sc_fifo_in_if<RGB_frame> >     rgb_pi;
  sc_port<sc_fifo_out_if<YCRCB_frame> > ycrcb_po;
};
```

*Figure 11-11.* Example of Port Interconnect Setup (1 of 3)

```
//FILE: YCRCB_Mixer.h
SC_MODULE(YCRCB_Mixer) {
  sc_port<sc_fifo_in_if<float> >          K_pi;
  sc_port<sc_fifo_in_if<YCRCB_frame> >  a_pi, b_pi;
  sc_port<sc_fifo_out_if<YCRCB_frame> > y_po;
};
```

*Figure 11-12.* Example of Port Interconnect Setup (2 of 3)

```
//FILE: VIDEO_Mixer.h
SC_MODULE(VIDEO_Mixer) {
  // ports
  sc_port<sc_fifo_in_if<YCRCB_frame> >  dvd_pi;
  sc_port<sc_fifo_out_if<YCRCB_frame> > video_po;
  sc_port<sc_fifo_in_if<MIXER_ctrl> >    control;
  sc_port<sc_fifo_out_if<MIXER_state> > status;
  // local channels
  sc_fifo<float>          K;
  sc_fifo<RGB_frame>    rgb_graphics;
  sc_fifo<YCRCB_frame> ycrcb_graphics;
  // constructor
  SC_HAS_PROCESS(VIDEO_Mixer);
  VIDEO_Mixer(sc_module_name nm);
  void Mixer_thread();
};
```

*Figure 11-13.* Example of Port Interconnect Setup (3 of 3)

Now, let's look at interconnection of the preceding modules using both named and positional syntaxes.

```
VIDEO_Mixer::VIDEO_Mixer(sc_module_name nm)
: sc_module(nm)
{
  // Instantiate
  Rgb2YCrCb Rgb2YCrCb_i("Rgb2YCrCb_i");
  YCRCB_Mixer YCRCB_Mixer_i("YCRCB_Mixer_i");
  // Connect
  Rgb2YCrCb_i.rgb_pi(rgb_graphics);
  Rgb2YCrCb_i.ycrcb_po(ycrcb_graphics);
  YCRCB_Mixer_i.K_pi(K);
  YCRCB_Mixer_I.a_pi(dvd_pi);
  YCRCB_Mixer_i.b_pi(ycrcb_graphics);
  YCRCB_Mixer_i.y_po(video_po);
};
```

*Figure 11-14.* Example of Port Interconnect by Name

Although slightly more code than the positional notation, the named port syntax is more robust, and tools exist to reduce the typing tedium.

```
VIDEO_Mixer::VIDEO_Mixer(sc_module_name nm)
: sc_module(nm)
{
  // Instantiate
  Rgb2YCrCb Rgb2YCrCb_i("Rgb2YCrCb_i");
  YCRCB_Mixer YCRCB_Mixer_i("YCRCB_Mixer_i");
  // Connect
  Rgb2YCrCb_i(rgb_graphics,ycrcb_graphics);
  YCRCB_Mixer_i(K,dvd_pi,ycrcb_graphics,video_po);
};
```

*Figure 11-15.* Example of Port Interconnect by Position

The problem with positional connectivity is keeping the ordering correct. In large designs, middle- and upper-level modules frequently have a large number of ports (potentially multiple 10's), and it is common to add or remove ports late in the design. Using a positional notation can quickly lead to debug problems. That is why we recommend avoiding the positional syntax entirely, and always using a named port approach.

GUIDELINE:   Whenever possible use the named port interconnection style.

How does it work? Whereas the complete details require an extensive investigation of the SystemC library code, we can provide a short answer. When the code instantiating an **sc_port** executes, the **operator**() is overloaded to take a channel object by reference and saves a pointer to that reference internally for later access by the port. Thus, we recall a port is an interface pointer to a channel that implements the interface.

## 11.6 Accessing Ports From Within a Process

Connecting ports between modules and channels is of no great value unless a process somewhere in the design can initiate activity over the channels. This section will show how to access ports from within a process. The **sc_port** overloads the C++ **operator->**(), which allows a simple syntax to access the referenced interface.

```
portname->method(optional_args);
```

*Figure 11-16.* Syntax of Port Access

Continuing the previous example, we now illustrate port access in action. In the following, **control** and **status** are the ports; whereas, **K** is a local channel instance. Notice use of the **operator->** when accessing ports.

```
void VIDEO_Mixer::Mixer_thread() {
  ...
  switch (control->read()) {
    case MOVIE: K.write(0.0f); break;
    case MENU:  K.write(1.0f); break;
    case FADE:  K.write(0.5f); break;
    default:    status->write(ERROR); break;
  }
  ...
}
```

*Figure 11-17.* Example of Port Access

A mnemonic may help here. P is for port and P is for pointer. When accessing channels through ports, always use the pointer method (i.e., ->).

## 11.7 Exercises

For the following exercises, use the samples provided in www.EklecticAlly/Book/.

**Exercise 11.1:** Examine, compile, and run the `video_mix` example.

**Exercise 11.2:** Examine, compile, and run the `equalizer_ex` example.

| Predefined Primitive Channels: Mutexes, FIFOs, & Signals | | | |
|---|---|---|---|
| Simulation Kernel | Threads & Methods | Channels & Interfaces | Data types: Logic, Integers, Fixed point |
| | Events, Sensitivity & Notifications | **Modules & Hierarchy** | |

# MORE ON PORTS
*Specialized & sc_export*

This chapter continues our discussion of ports as we go beyond the basics and explore intermediate concepts. We start out with a look at some standard interfaces that can be used to build ports. Next, we discuss built-in specialized ports and their conveniences, especially with regard to static sensitivity. Finally, we present the concept of port arrays, and finish the chapter with a new type of port, **sc_export**, that is new to SystemC version 2.1.

## 12.1 Standard Interfaces

SystemC provides a variety of standard interfaces that go hand in hand with the built-in channels discussed previously. This section describes these interfaces. This presentation is a more precise definition of the syntax, and this material provides a basis for creating custom channels that will be discussed in Chapter 13.

### 12.1.1  sc_fifo interfaces

Two interfaces, **sc_fifo_in_if**<>, and  **sc_fifo_out_if**<>, are provided for **sc_fifo**<>. Together these interfaces provide all of the methods implemented by **sc_fifo**<>. In fact, the interfaces were defined prior to the creation of the channel. The channel simply becomes the place to implement the interfaces and holds the data implied by a FIFO functionality.

The interface, **sc_fifo_out_if**<>, partially shown in *Figure 12-1*, provides all the methods for output from a module into an **sc_fifo**<>. The module pushes data onto the FIFO using **write**() or **nb_write**(). The **num_free**() indicates how many locations are free, if any. The **data_read_event**() method may be used to dynamically wait for a location to be freed. We've discussed all of these methods in Chapter 8, Basic Channels when we discussed the **sc_fifo<>** channel.

Notice in the following figure that the interface itself is templated on a class name just like the corresponding channel.

```
// Definition of sc_fifo<T> output interface
template <class T>
struct sc_fifo_out_if: virtual public sc_interface {
  virtual void write(const T& ) = 0;
  virtual bool nb_write(const T& ) = 0;
  virtual int num_free() const = 0;
  virtual const sc_event&
                  data_read_event() const = 0;
};
```

*Figure 12-1.* sc_fifo Output Interface Definitions—Partial

The other interface, **sc_fifo_in_if<>**, provides all the methods for input to a module from an **sc_fifo<>**. The module pulls data from the FIFO using **read()** or **nb_read()**. The **num_available()** indicates how many locations are occupied, if any. The **data_written_event()** method may be used to dynamically wait for a new value to become available.

Again, we discussed all of these methods in Chapter 8, when we discussed the **sc_fifo<>** channel.

Here is the corresponding portion of the interface definition from the SystemC library:

```
// Definition of sc_fifo<T> input interface
template <class T>
struct sc_fifo_in_if: virtual public sc_interface {
  virtual void read( T& ) = 0;
  virtual T read() = 0;
  virtual bool nb_read( T& ) = 0;
  virtual int num_available() const = 0;
  virtual const sc_event&
          data_written_event() const = 0;
};
```

*Figure 12-2.* sc_fifo Input Interface Definitions—Partial

Something interesting to notice about the **sc_fifo** interfaces is that if you simply use the **read()** and **write()** methods in your module and do not rely on the other methods, then your use of these interfaces is very compatible with the corresponding **sc_signal<>** interfaces, which we will discuss next. In other words, it is almost conceivable to simply swap out the interfaces; however, doing so would be dangerous. Remember **sc_fifo::read()** and **sc_fifo::write()** block waiting for the FIFO to empty; whereas, **sc_signal::read()** and **sc_signal::write()** are non-blocking.

### 12.1.2 sc_signal interfaces

Similar to **sc_fifo<>**, two interfaces, **sc_signal_in_if<>**, and **sc_signal_out_if<>**, are provided for **sc_signal<>**. In addition, a third interface, **sc_signal_inout_if<>**, provides bi-directional capability. Together these interfaces provide all of the methods provided by **sc_signal<>**. Again, the interfaces were defined prior to the creation of the channel. The channel simply becomes the place to implement the interfaces and provides the request-update behavior implied for a signal.

The request portion of the behavior is provided by the **sc_signal_inout_if<>** interface.

Here is a portion of the interface definition as provided in the SystemC library:

```
// Definition of sc_signal<T> input/output interface
template <class T>
struct sc_signal_inout_if: public sc_signal_in_if<T>
{
  virtual void write( const T& ) = 0;
};
#define sc_signal_out_if sc_signal_inout_if
```

*Figure 12-3.* sc_signal Input Interface Definitions - Partial

There are two rather interesting things to notice in the preceding interface. First, **sc_signal_out_if**<> is defined as a synonym to **sc_signal_inout_if**<>. This definition lets a module read the value of an output signal directly rather than being forced to keep a local copy in the manner that VHDL requires.

The other feature to notice is how **sc_signal_inout_if**<> inherits the input behavior from **sc_signal_in_if**<>. In fact, the **sc_signal**<> channel inherits directly from this channel, **sc_signal_inout_if**<>, rather than the other two interfaces.

The update portion of the behavior is provided as a result of a call to **request_update**() that is provided indirectly as a result of a call from **sc_signal::write**(). The update is implemented with the protected **sc_signal::update**() method call. The **sc_signal_in_if**<> interface provides access to the results through **sc_signal::read**().

```
// Definition of sc_signal<T> input interface
template <class T>
struct sc_signal_in_if: virtual public sc_interface {
  virtual const sc_event&
                  value_changed_event() const = 0;
  virtual const T& read() const = 0;
  virtual const T& get_data_ref() const = 0;
  virtual bool event() const = 0;
};
```

*Figure 12-4.* sc_signal Input Interface Definitions

The first road from San Marcos, a 15-mile wagon track, took most of the day to traverse. It ran across steep hills and low water crossings, and through pastures, entailing numerous stops to open and close gates. When the automobile came along, writes Dorothy Wimberley Kerbow in *Wimberley, Historic Belle of the Blanco*, "A few brave drivers drove . . . over the wagon road, changing flat tires frequently, and losing their religion equally." In 1921 a caravan of 100 cars celebrated the opening of a new gravel road that considerably shortened the trip. This attracted people from Houston and San Antonio who built summer homes. The gravel road was paved in the 1940s, again shortening the trip and increasing the town's popularity. After World War II Wimberley's climate, scenery, waterways, and growing art colony increased its popularity as a resort community. Market Days is a popular Wimberley attraction. This huge marketplace, a combined craft show, antique mall, and flea market, began in 1964 as a once-a-year event. Now, under sponsorship of the Lions Club, it is held the first Saturday of each month, April to December.

Besides the walk described here, two additional Wimberley walks present spectacular views.

North of town, off RR 2325 in Woodcreek, a walk up Old Baldy's 218 steps is rewarded with a panorama of the valley. Ask for directions. This is not recommended during hot weather! Mt. Baldy is one of the Twin Sister peaks, a Wimberley Valley landmark. *Wimberley Mountaineer* editor Ernest H. Baxter named them Mount Alberta (1,182 feet) and Mount Edith (1,219 feet) for two of rancher Austin A. Lowery's daughters, who were not twins. Steps date from 1950, when John E. Harris and Odessa Farris laid a dance floor and installed a juke box on the summit when Cactus Pryor's "On Top of Old Baldy" was popular. Since the mountain didn't have a cedar covering like its twin, its name changed. In the mid-1990s Trinity Chapel bought Mt. Baldy and calls it Prayer Mountain.

Also off RR 2325, on the left just past the high school, the EmilyAnn Theatre offers a short uphill walk on a soft juniper mulch trail. Along the way you pass the theater, Gluttons Glade with its picnic tables, Puppets Post, Jugglers Junction, the Trothing Tree, and Poet's Podium. At the top you get a direct view of Mt. Baldy. More views await as you descend on the other side of the hill. Wimberley High School has run an accredited six-week summer Shakespeare program since 1990. It attracts students from all over the state. Performance were in the courtyard of the school until the community built the EmilyAnn 1998 as a tribute to a 16-year-old high school drama student who lost her in an auto accident.

# WIMBERLEY WALK

Square, historic homes, millrace, and cemetery. 2.5 miles. Unfortunately, except around the square, roads have narrow or rough shoulders suitable for walking but not for wheelchairs and doubtful for strollers. Park around the Square. Public restrooms at the Visitor Center (Step 6) on RR 12 north of the traffic light, open Monday to Saturday 9 a.m. to 4 p.m. and Sunday 1 to 4 p.m. There is a variety of restaurants on the square and around town.

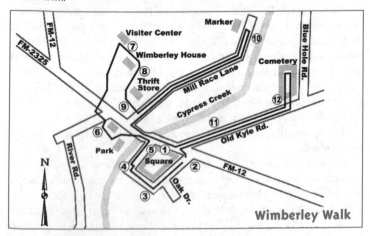

Wimberley Walk

## ❶ On the square, face the complex of buildings in the center.

As you will surmise immediately, the "square" is not square. It once was—more or less—and the name stuck. The building in the center sits on the site of the Pyland home and blacksmith shop. There is a historic marker half hidden by the cafe tables out front. Sidney Pyland and brother John Will married Pleasant Wimberley's granddaughters. Pyland moved his blacksmith shop to Marcos in 1910.

1933 George and Christine Burdett moved to Wimberley against the f their banker, who warned, "Nobody in Wimberley has ever made any imberley is just like something that fell off the back end of a garbage followed their own instincts, however, and opened a small store at ion.

 ll, they bought the west side of the square and built Burdett's anch House Cafe. Christine collected rocks of many kinds several of Wimberley's finest masons to build "the most ." The Burdetts lived above the store and ran a hotel was a feed store and ice house. Although they sold and continued in business until destroyed by fire in 1965.

### 12.1.3   sc_mutex and sc_semaphore interfaces

The two channels, **sc_mutex** and **sc_semaphore**, also provide interfaces for use with ports. It is interesting to note that neither interface provides any event methods for sensitivity. If you require event sensitivity, you must write your own channels and interfaces as discussed in the next chapter.

```
// Definition of sc_mutex_if interface
struct sc_mutex_if: virtual public sc_interface {
  virtual int lock() = 0;
  virtual int trylock() = 0;
  virtual int unlock() = 0;
};
```

*Figure 12-5.* sc_mutex Interface Definitions

```
// Definition of sc_semaphore_if interface
struct sc_semaphore_if : virtual public sc_interface
{
  virtual int wait() = 0;
  virtual int trywait() = 0;
  virtual int post() = 0;
  virtual int get_value() const = 0;
};
```

*Figure 12-6.* sc_semaphore Interface Definitions

## 12.2 Static Sensitivity Revisited

Recall from Chapter 7, Concurrency, that processes can be made sensitive to events. Also recall from Chapter 8, Basic Channels, that standard channels often provide methods that provide references to events (e.g., **sc_fifo::data_written_event()**). Since ports are defined on interfaces to channels, it is only natural to want sensitivity to events defined on those channels.

For example, it might be nice to create an **SC_METHOD** statically sensitive to the **data_written_event** or perhaps monitor an **sc_signal** for any change in the data using the **value_changed_event**. You might even want to monitor a subset of possible events such as a positive edge transition (i.e., **false** to **true**) on an **sc_signal<bool>**.

This approach has a minor difficulty. Ports are pointers that become initialized during elaboration, and they are undefined at the point in time when the **sensitive** method needs to know about them. SystemC provides a solution for this difficulty in the form of a special class, the **sc_event_finder**.

The **sc_event_finder** lets the determination of the actual event be deferred until after elaboration. Unfortunately, the **sc_event_finder** has a minor complication. An **sc_event_finder** must be defined for each event defined by the interface. Thus, it is commonplace to define template specializations of port/interface combinations to instantiate a port, and include an **sc_event_finder** in the specialized class.

Suppose you want to create a port with sensitivity to the positive edge event of a Boolean signal port using the **sc_signal_in_if<bool>::posedge_event()** member function as shown in the following example:

```
struct ea_port_sc_signal_in_if_bool
: public sc_port<sc_signal_in_if<bool>,1>
{

  typedef sc_signal_in_if<bool> if_type;//typing aid
  sc_event_finder& ef_posedge_event() const {
    return *new sc_event_finder_t<if_type>(
                  *this,
                  &if_type::posedge_event
                );
  }
};
```

*Figure 12-7.* Example of Event Finder

Let's examine the preceding example. First, our custom event finder is inheriting from a port specification on the second line. Second, to save some typing, we've created a **typedef** called if_type, which refers to the interface specialization. The new method, ef_posedge_event()[32] creates a new **event_finder** object and returns a reference to it. The constructor for an **sc_event_finder** takes two arguments: a reference to the port being found (*this), and a reference to the member function,

---

[32] The prefix ef_ is a convention. Some groups might prefer a suffix, _ef. In any case a convention should be adopted.

(**posedge_event**()) that returns a reference to the event of interest. The preceding example returns a reference to the event finder, which can be used by **sensitive**.

Now, the preceding specialization may be used as follows:

```
SC_MODULE(my_module) {
  ea_port_sc_signal_in_if_bool my_p;
  ...
  SC_CTOR(...) {
    SC_METHOD(my_method);
    sensitive << my_p.ef_posedge_event();
  }
  void my_method();
  ...
};
```

*Figure 12-8.* Example of Event Finder Use

A related and useful concept for sensitivity lists in SystemC is the ability to be sensitive to a port. The idea is that a process, typically an **SC_METHOD**, is concerned with any change on an input port. Obviously, this may be coded directly.

As a syntactical simplification, SystemC also allows specifying a port name if and only if the associated interface provides a method called **default_event**() that returns a reference to an **sc_event**. The standard interfaces for **sc_signal** and **sc_buffer** provide this method. If you design your own interfaces, you will need to supply this method yourself.

## 12.3 Specialized Ports

Event finders are not particularly difficult to code; however, they are additional coding. To reduce that burden, SystemC provides a set of template specializations that provide port definitions on the standard interfaces and include the appropriate event finders.

It is important to know the port specializations for two reasons. First, you will doubtless have need for the common event finders at some point. Second, you will encounter their use in code from other engineers.

Let's take a look at the syntax of FIFO specializations:

```
// sc_port<sc_fifo_in_if<T> >
sc_fifo_in<T> name_fifo_ip;
sensitive << name_fifo_ip.data_written();
value = name_fifo_ip.read();
name_fifo_ip.read(value);
if (name_fifo_ip.nb_read(value))...
if (name_fifo_ip.num_available())...
wait(name_fifo_ip.data_written_event());


// sc_port<sc_fifo_out_if<T> >
sc_fifo_out<T> name_fifo_op;
sensitive << name_fifo_op.data_read();
name_fifo_op.write(value);
if (name_fifo_op.nb_write(value))...
if (name_fifo_op.num_free())...
wait(name_fifo_op.data_read_event());
```

Don't use.
Prefer the
-> syntax.

*Figure 12-9.* Syntax of FIFO Port Specializations

These specializations have a minor downside that has to do with how ports are to be referenced. Notice in the following syntax figures that methods such as **read**() are defined. Recall from the last chapter that processes invoke port methods using **operator**->(). With specialized ports, you may also use dot (e.g., my_sig.**read**()). This syntax has the unfortunate effect of creating bad habits that could cause you to stumble later.

You may still use the pointer methods in processes. With exception to the new **sc_event_finder** methods and initialization, we recommend you try to stick to the pointer form whenever possible.

GUIDELINE: Use dot (.) in the elaboration section of the code, but keep to arrow (->) in processes.

This style will help you differentiate port accesses from local channel accesses and reduce confusion.

Let's look at an example using the FIFO port specializations using the guideline:

```
// Equalizer.h
SC_MODULE(Equalizer) {
  sc_fifo_in<double> raw_fifo_ip;
  sc_fifo_out<double> equalized_fifo_op;
  void equalizer_thread();
  SC_CTOR(Equalizer) {
    SC_THREAD(equalizer_thread);
      sensitive << raw_fifo_ip.data_written();
  }
};
```

Only available in port specialization.

```
// Equalizer.cpp
void Equalizer::equalizer_thread() {
  for(;;) {
    double sample;
    wait();
    raw_fifo_ip->nb_read(sample);
    ... /* process data */
    equalized_fifo_op->write(result);
  }//endforever
}
```

*Figure 12-10.* Example Using FIFO Port Specializations

Now we'll examine specialized ports for evaluate-update channels such as **sc_signal**<>. We left out the obvious duplication of member functions such as **read**() that are better handled using the arrow (->) operator.

```
// sc_port<sc_signal_in_if<T> >
sc_in<T> name_sig_ip;
sensitive << name_sig_ip.value_changed();

// Additional sc_in specializations...
sc_in<bool> name_bool_sig_ip;
sc_in<sc_logic> name_log_sig_ip;
sensitive << name_sig_ip.pos();
sensitive << name_sig_ip.neg();


// sc_port<sc_signal_out_if<T> >
sc_inout<T> name_sig_op;
sensitive << name_sig_op.value_changed();
sc_inout_resolved<N> name_rsig_op;
sc_inout_rv<N> name_rsig_op;
sc_out<T> name_rsig_op;
sc_out_resolved<T> name_rsig_op;
sc_out_rv<T> name_rsig_op;
// everything under sc_in<T> plus the following...
name_sig_op.initialize(value);
name_sig_op = value; // <-- DON'T USE!!!
```

*Figure 12-11.* Syntax of Signal Port Specializations

In the preceding syntax, there are several features to note. First, there is the **initialize**() method. This method may be used at elaboration to establish the initial values of signal ports. This approach is handy to model start-up conditions properly rather than wait a delta-cycle and mess with synchronizing processes at the start.

We also included one additional syntax that we especially don't like, the assignment operator (=), on the last line. When used, this operator can be especially confusing because unless you realize the name on the left is a signal port, the behavior will appear to be bizarre. Remember that signals have an evaluate-update behavior, which is quite different from ordinary assignment.

GUIDELINE: To avoid confusion, never use the assignment operator shortcut with **sc_signal** or **sc_port**<sc_signal>. Instead, prefer the **write**() method.

Let's look at an example using signal port specializations. This example is a typical hardware block, a 32-bit linear feedback shift register (LFSR) commonly used with built-in self-test (BIST). Notice the use of **pos**() and **initialize**().

```
//FILE: LFSR_ex.h
#ifndef LFSR_EX_H
#define LFSR_EX_H
SC_MODULE(LFSR_ex) {
  // Ports
  sc_in<bool> sample;
  sc_out<sc_int<32> > signature;
  sc_in<bool> clock;
  sc_in<bool> reset;
  // Constructor
  SC_CTOR(LFSR_ex) {
    // Register process
    SC_METHOD(LFSR_ex_method);
    sensitive << clock.pos() << reset;
    signature.initialize(0);
  }
  // Process declarations & Local data
  void LFSR_ex_method();
  sc_int<32> LFSR_reg;
};
#endif
```

*Figure 12-12.* Example of Signal Port Specializations - Header

```
//FILE: LFSR_ex.cpp
#include <systemc.h>
#include "LFSR.h"
void LFSR_ex::LFSR_ex_method() {
  if (reset->read() == true) {
    LFSR_reg = 0;
    signature->write(LFSR_reg);
  } else {
    bool lsb =LFSR_reg[31]^LFSR_reg[25]^LFSR_reg[22]
              ^LFSR_reg[21]^LFSR_reg[15]^LFSR_reg[11]
              ^LFSR_reg[10]^LFSR_reg[ 9]^LFSR_reg[ 7]
              ^LFSR_reg[ 6]^LFSR_reg[ 4]^LFSR_reg[ 3]
              ^LFSR_reg[ 1]^LFSR_reg[ 0]
              ^ sample->read();
    LFSR_reg.range(31,1) = LFSR_reg.range(30,0);
    LFSR_reg[0] = lsb;
    signature->write(LFSR_reg);
  }//endif
}
```

*Figure 12-13.* Example of Signal Port Specializations - Implementation

## 12.4 The sc_port<> Array

The **sc_port**<> provides a second parameter we have not yet discussed, the array size parameter. The idea is to provide for the possibility of a number of like-defined ports. This is referred to as a multi-port or port array.

For example, a communications system might have a number of T1 interfaces all with the same connectivity. Another example might be an abstract hierarchical communications channel that may have any number of devices connected to it. The full **sc_port** syntax follows.

```
sc_port <interface[,N]> portname;
// N=0..MAX Default N=1
```

*Figure 12-14.* Syntax of sc_port<> Declaration Complete

For $N \neq 0$, then precisely **N** channels must be connected to the port. The case where **N** = 0 is a special case that allows an almost unlimited number of ports. In other words, you may connect any number of channels to the port.

An example with a drawing may help with understanding.

## Multi-Ports

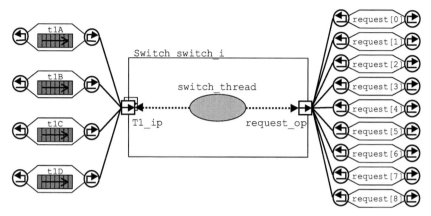

*Figure 12-15.* Illustration of sc_port<> Array Connectivity

Here is the header code for the switch:

```
//FILE: Switch.h
SC_MODULE(Switch) {
  sc_port<sc_fifo_in_if<int>,4> T1_ip;
  sc_port<sc_signal_out_if<bool>,0> request_op;
  ...
};
```

*Figure 12-16.* Example of sc_port<> Array Declaration

Channels are connected to port arrays the same way ordinary ports are connected, except port arrays have more than one connection. In fact, the basic port syntax simply relies on the default that **N** = 1. When **N** > 1, each connection is assigned a position in the array on a first-connected first-position basis.

Here is the corresponding example for the connections:

```
//FILE: Board.h
#include "Switch.h"          From preceding
SC_MODULE(Board) {           example.
  const unsigned REQS;
  Switch switch_i;
  sc_fifo<int> t1A, t1B, t1C, t1D;
  sc_signal<bool> request[9];
  SC_CTOR(Board): switch_i("switch_i")
  {
    // Connect 4 T1 channels to the switch
    // input T1 ports
    switch_i.T1_ip(t1A); switch_i.T1_ip(t1B);
    switch_i.T1_ip(t1C); switch_i.T1_ip(t1D);
    // Connect 9 request channels to the
    // switch request output ports
    for (unsigned i=0;i!=9;i++) {
      switch_i.request_op(request[i]);
    }//endfor
    ...
  }//end constructor
  ...
};
```

*Figure 12-17.* Example of sc_port<> Array Connections

The preceding example illustrates several things. First, a fixed port array of size 4 is connected directly to four FIFO channels. Second, an unbounded array is connected to an array of channels using a for-loop.

Access to port arrays from within a process is accomplished using an array syntax. This class also provides a method, **size**(), that may be used to examine the declared port size. This method is useful for situations where the array bounds are unknown (i.e., $N = 0$).

Here is the code implementing the process accessing the multi-ports:

```cpp
//FILE: Switch.cpp
void Switch::switch_thread() {
  // Initialize requests
  for (unsigned i=0;i!=request_op.size();i++) {
    request_op[i]->write(true);
  }//endfor
  // Startup after first port is activated
  wait(T1_ip[0]->data_written_event()
      |T1_ip[1]->data_written_event()
      |T1_ip[2]->data_written_event()
      |T1_ip[3]->data_written_event()
  );
  for(;;) {
    for (unsigned i=0;i!=T1_ip.size();i++) {
      // Process each port...
      int value = T1_ip[i]->read();
    }//endfor
  }//endforever
}//end Switch::switch_thread
```

*Figure 12-18.* Example of sc_port<> Array Access

Notice that the **size**() method requires the dot operator because it's defined in the specialized port class (e.g., request_op or T1_ip) rather than in the external channel (e.g., request[i] or t1A, t1B, t1C, t1D). On the other hand, port access to the channel uses the arrow operator as would be expected.

One current syntactical downside to **wait**() syntax may be seen in the preceding syntax. If you need to use "any event in the attached channels," current syntax requires an explicit listing.

Version 2.1 offers an alternate possibility with dynamic threads (specifically, fork/join statements). One could create and launch a separate thread to monitor each port and provide communication back via a shared local variable. We will examine this feature in Chapter 14, Advanced Topics.

## 12.5 SystemC Exports

SystemC version 2.1[33] provides a new type of port called the **sc_export**. This port is similar to standard ports in that the declaration syntax is defined on an interface, but this port differs in connectivity. The idea of an **sc_export** is to move the channel inside the defining module, and use the port externally as though it were a channel. The following figure illustrates this concept:

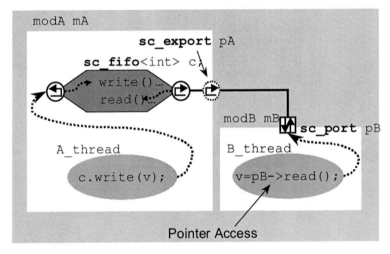

*Figure 12-19.* How sc_export Works

Contrast this concept with Figure 11-9 where we originally investigated ports.

The observant programmer might ask, why use **sc_export** at all? After all, one could just access the internal **sc_channel** instance name directly using a hierarchical access. That approach works only if the interior channel is publicly accessible. For an IP provider, it may be desirable to export only specific channels and keep everything else private. Thus, **sc_export** allows control over the interface.

Another reason for using **sc_export** is to provide multiple interfaces at the top level. Without **sc_export**, we are left to using the hierarchical channel, and that allows for only a single top-level set of interfaces.

---

[33] If you do not have version 2.1 available, don't fret. You can access much of the functionality using hierarchical channels.

With export, each **sc_export** contains a specific interface. Since connection is not required, **sc_export** also allows creation of "hidden" interfaces. For example, a debug or test interface might be used internally by an IP provider, but not documented for the end user. Additionally, it allows a development team or IP provider to develop an "instrumentation model" that can be used extensively during architectural exploration and then dropped during regression runs when visibility is less needed and performance is key.

The syntax follows **sc_port**, but without the multi-port possibility.

```
sc_export<interface> portname;
```

*Figure 12-20.* Syntax of sc_export<> Declaration

Connectivity to an **sc_export** requires some slight changes since the channel connections have now moved inside the module. Thus, we have:

```
SC_MODULE(modulename) {
  sc_export<interface> portname;
  channel cinstance;
  SC_CTOR(modulename) {
    portname(cinstance);
  }
};
```

*Figure 12-21.* Syntax sc_export<> Internal Binding to Channel

Let's look at a simple example. This example provides a process that is toggling an internal signal periodically. The **sc_export** in this case is simply the toggled signal. For compactness, this example includes the entire module definition in the header.

```
SC_MODULE(clock_gen) {
  sc_export<sc_signal<bool> > clock_xp;
  sc_signal<bool> oscillator;
  SC_CTOR(clock_gen) {
    SC_METHOD(clock_method);
    clock_xp(oscillator); // connect sc_signal
                          // channel
                          // to export clock_xp
    oscillator.write(false);
  }
  void clock_method() {
    oscillator.write(!oscillator.read());
    next_trigger(10,SC_NS);
  }
};
```

*Figure 12-22.* Example of Simple sc_export<> Declaration

To use the above **sc_export**, we provide the corresponding instantiation of this simple module.

```
#include "clock_gen.h"
...
clock_gen clock_gen_i;
collision_detector collision_detector_i;
// Connect clock
collision_detector_i.clock(clock_gen_i.clock_xp);
...
```

*Figure 12-23.* Example of Simple sc_export<> Instantiation

Another powerful possibility with **sc_export** is to let interfaces be passed up the design hierarchy as illustrated in the next figure.

The following figure illustrates this idea:

## Hierarchy with sc_exports

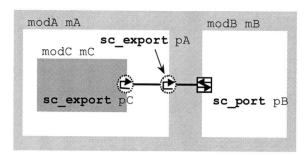

*Figure 12-24.* sc_export Used with Hierarchy

Just like **sc_port**, the **sc_export** can be bound directly to another **sc_export** in the hierarchy. Here is how to accomplish this binding:

```
SC_MODULE(modulename) {
  sc_export<interface> xportname;
  module minstance;
  SC_CTOR(modulename), minstance("minstance") {
    xportname(minstance.subxport);
  }
};
```

*Figure 12-25.* Syntax of sc_export<> Binding to sc_export<>

The **sc_export** allows easy access to internal channels for debug. One nice aspect of **sc_export** is the lack of a requirement for a connection. By contrast, an **sc_port** requires a connection.

**sc_export** has some caveats that may not be obvious. First, it is not possible to use **sc_export** in a static sensitivity list. On the other hand, you can access the interface via **operator->()**. Thus, one can use **wait**(*xportname->event()*) on suitably defined interfaces accessed within an **SC_THREAD**.

Second, as previously mentioned, it is not possible to have an array of **sc_export** in the same manner as **sc_port**. On the other hand, suitable channels may allow multiple connections, which may make this issue moot.

The following is an example of how an **sc_export** might be used to model a complex bus including an arbiter to be provided as an IP component. First let's look at the customer view:

**Customer View**

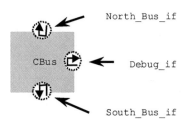

*Figure 12-26.* Example of Customer View of IP

This would be provided as the following header file:

```
//CBus.h
#ifndef CBUS_H
#define CBUS_H
#include "CBus_if.h"
class North_bus; // Forward declarations
class South_bus;
class Debug_bus;
class CBus_rtl;
SC_MODULE(CBus) {
  sc_export<CBus_North_if> north_p;
  sc_export<CBus_South_if> south_p;
  SC_CTOR(CBus);
private:
  North_bus* nbus_ci;    South_bus* sbus_ci;
  Debug_bus* debug_ci;   CBus_rtl*  rtl_i;
};
#endif
```

*Figure 12-27.* Example of sc_export Applied to a Bus

Notice how the preceding is independent of the implementation, and the end user is not compelled to hook up either bus. In addition, the debug interface is not provided in this example header. Here is the implementation view:

**Vendor View**

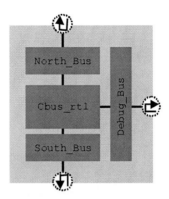

*Figure 12-28.* Example of Vendor View of IP

Here is the implementation code, which may be kept private:

```
//FILE: CBus.cpp
#include "CBus.h"
#include "North_bus.h"
#include "South_bus.h"
#include "Debug_bus.h"
#include "CBus_rtl_bus.h"
CBus::CBus(sc_module_name nm): sc_module(nm) {
  // Local instances
  nbus_ci  = new North_bus("nbus_ci");
  sbus_ci  = new South_bus("sbus_ci");
  debug_ci = new Debug_bus("debug_ci");
  rtl_i    = new CBus_rtl("rtl_i");
  // Export connectivity
  north_p(nbus_ci);
  south_p(sbus_ci);
  // Implementation connectivity
  ...
}
```

*Figure 12-29.* Example of sc_export Applied to a Bus Constructor

In the preceding code, notice the debug interface is not provided to the customer. This would be an optional aspect of the IP.

## 12.6 Connectivity Revisited

Let's review port connectivity. The following diagram is copied from the previous chapter. It should become second nature to understand how to accomplish all the connections illustrated.

# Port Connections

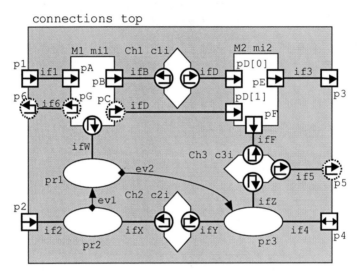

*Figure 12-30.* Connectivity Possibilities

All of the possible connections are illustrated in this one figure. This figure is a handy reference when reviewing the SystemC connection rules, which are listed below:

1. Processes may communicate with other processes in the same module using channels. For example, process `pr2` to process `pr3` via interface `ifX` on channel `c2i`.

2. Processes may communicate with other processes in the same module using events to synchronize exchanges of information through data variables instantiated at the module level (e.g., within the module class definition). For example, process `pr2` to process `pr1` via event `ev1`.

3. Processes may communicate with processes upwards in the design hierarchy using the interfaces accessed via `sc_port`. For example, process `pr2` via port `p2` using interface `if2`.

4. Processes may communicate with processes in sub-module instances via interfaces to channels connected to the sub-module ports. For example, process `pr3` to module `mi2` via interface `ifZ` on channel `c3i`.

5. **sc_export**s may connect via interfaces to local channels. For example, port `p5` to channel `c3i` using interface `if5`.

6. **sc_port**s may connect directly to **sc_port**s of sub-modules. For example, port `p1` is connected to port `pA` of sub-module `mi1`

7. **sc_export**s may connect directly to **sc_export**s of sub-modules. For example, port `p6` is directly connected to port `pG` of sub-module `mi1`.

8. **sc_port**s may connect indirectly to processes by letting the processes access the interface. This is just a process accessing a port described previously. For example process `pr1` communicates with sub-module `mi1` through interface `ifW`

9. **sc_port** arrays may be used to create multiple ports using the same interface. For example, `pD[0]` and `pD[1]` of sub-module `mi2` constitute a port array.

Finally, we present an equivalent diagram to the preceding. In this diagram, channels appear as slightly thickened lines. The **sc_port**s are represented with a square containing a circle to indicate the presence of an interface. This style is often used to simplify the schematic representation, but at the expense of slightly hiding the underlying functionality. In the next chapter, we will investigate more complex channels known as hierarchical channels.

## Hidden Channels

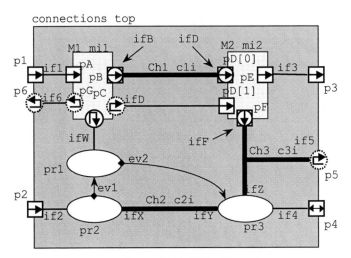

*Figure 12-31.* Hidden Channels

## 12.7 Exercises

For the following exercises, use the samples provided at www.EklecticAlly.com/Book/.

**Exercise 12.1:** Examine, compile, and run the static_sensitivity example.

**Exercise 12.2:** Examine, compile, and run the connections example. See if you can identify all the connections shown in the figures in this chapter.

| | | Channels & Interfaces | |
|---|---|---|---|
| | | | |

Predefined Primitive Channels: Mutexes, FIFOs, & Signals

| Simulation Kernel | Threads & Methods | Channels & Interfaces | Data types: Logic, Integers, Fixed point |
|---|---|---|---|
| | Events, Sensitivity & Notifications | Modules & Hierarchy | |

Chapter   13

# CUSTOM CHANNELS AND DATA
## *Primitive & Hierarchical*

By now, we've covered much of the syntax of SystemC. Now we will focus on some of the more abstract concepts from which SystemC derives much of its power. This chapter illustrates how to create custom channels of a variety of types including: primitive channels, custom signals, custom hierarchical channels, and custom adapters. Of these, custom signals and adaptors are probably the most commonly encountered.

## 13.1 A Review of Channels and Interfaces

Modules control and process data. Channels implement communications between modules. Interfaces provide a mechanism to allow independence of modules from the mechanisms of communication, channels.

Channels come in two flavors: primitive and hierarchical. The basic premise of a channel is a class that inherits from an interface. The interface makes a channel usable with ports. In addition, channels must inherit either from **sc_prim_channel** or **sc_channel**. This distinction in these latter two base classes is one of distinct capabilities and features. In other words, **sc_prim_channel** has capabilities not present in **sc_channel** and visa versa.

Primitive channels are intended to provide very simple and fast communications. They contain no hierarchy, no ports, and no **SC_METHOD**s or **SC_THREAD**s. Primitive channels have the ability to implement the evaluate-update paradigm. We will discuss how to do this in Section 13.2.

By contrast, hierarchical channels may access ports, they can have processes and contain hierarchy as the name suggests. In fact, hierarchical channels are really just modules that implement one or more interfaces. Hierarchical channels are intended to model complex communications buses such as PCI, HyperTransport™, or AMBA™. We will briefly look at custom hierarchical channels in Section 13.4.

Channels are important in SystemC because they enable several concepts:

- Appropriate channels enable safe communication between processes
- Channels in conjunction with ports clarify the relationships of communication (producer vs. consumer)

Interfaces are important in SystemC because they enable the separation of communication from processing.

Let us now proceed and examine various channel designs.

## 13.2 The Interrupt, a Custom Primitive Channel

We discussed events in Chapter 7, Concurrency, and we saw how processes may use events to coordinate activities. We introduced hierarchy and ports in Chapter 10, Structure. We left the question of how we could provide a simple event or interrupt between processes located in different modules unanswered.

One approach might take a channel that has an event and simply use the side effect like `sc_signal<bool>`. However, using side effects is unsatisfying. Let us see how we might create a custom channel just for this purpose.

A proper channel must have an interface or interfaces to implement. The ideal interface provides only the methods required for a particular purpose. For our channel, we'll create two interfaces. One interface for sending events is called `ea_interrupt_gen_if`, and another interface for receiving events is called `ea_interrupt_evt_if`. To allow and simplify use in static sensitivity lists, we'll specify a `default_event`.

The interfaces are shown in the *Figure 13-1*. Notice that interfaces are required to inherit from the `sc_interface` base class. Also, the `default_event` has a specific calling signature to be properly recognized by `sensitive`.

```
struct ea_interrupt_gen_if: public sc_interface {
  virtual void notify() = 0;
  virtual void notify(sc_time t) = 0;
};
```

```
struct ea_interrupt_evt_if: public sc_interface {
  virtual const sc_event& default_event() const = 0;
};
```

*Figure 13-1.* Example of Custom Channel Interface

Next, we look at the implementation of the primitive channel shown in *Figure 13-2*. The implementation has four features of interest.

First, the channel must inherit from **sc_prim_channel** and the two interfaces defined previously.

Second, the constructor for the channel has similar requirements to an **sc_module**; it must construct the base class **sc_prim_channel**. This channel is provided in only one format, which is a format that requires an instance name string.

Third, the channel must implement the methods compelled by the interfaces from which it inherits.

Fourth, we specify a private implementation of the copy constructor to prevent its use. Simply put, a channel should never be copied. This feature of the implementation provides a compile-time error if copying is attempted.

```
#include "ea_interrupt_evt_if.h"
#include "ea_interrupt_gen_if.h"

struct ea_interrupt
: public sc_prim_channel
, public ea_interrupt_evt_if
, public ea_interrupt_gen_if
{
  // Constructors
  explicit ea_interrupt()
  : sc_prim_channel(
      sc_gen_unique_name("ea_interrupt")) {}
  explicit ea_interrupt(sc_module_name nm)
  : sc_prim_channel(nm) {}
  // Methods
  void notify() { m_interrupt.notify(); }
  void notify(sc_time t) { m_interrupt.notify(t); }
  const sc_event& default_event() const
    { return m_interrupt; }
private:
  sc_event m_interrupt;
  // Copy constructor so compiler won't create one
  ea_interrupt( const ea_interrupt& ) {}
};
```

*Figure 13-2.* Example of Custom Interface Implementation

## 13.3 The Packet, a Custom Data Type for SystemC

Creating custom primitive channels is not very common; however, creating an **sc_signal** channel or an **sc_fifo** is. SystemC defines all the necessary features for both of these channels when used with built-in data types. For custom data types, SystemC requires you to define several methods for your data type.

The reasons for the required methods are easy to understand. For instance, both channels support **read** and **write** methods, which involve copying. For this reason, SystemC requires the definition of the assignment operator (i.e., **operator=()**). Also, **sc_signal** supports the method **value_changed_event()**, which implies the use of comparison. In this case, SystemC requires the definition of the equality operator (i.e., **operator**==()).

Finally, there are two other methods required by SystemC, streaming output (i.e., **ostream**& **operator**<<()) and **sc_trace**. Streaming output allows for a pleasant printout of your data structure during debug. The trace function allows all or parts of your data type to be used with the SystemC trace facility, and this function enables viewing of trace data with a waveform viewer. We'll cover waveform data tracing in the next chapter.

Consider the following C/C++ custom data type, which might be used for PCI-X transactions:

```
struct pcix_trans {
    int devnum;
    int addr;
    int attr1;
    int attr2;
    int cmd;
    int data[8];
    bool done;
};
```

*Figure 13-3.* Example User Defined Data Type

This structure or record contains all the information necessary to carry on a PCI-X transaction; however, it is not usable with an **sc_signal** channel or an **sc_fifo**. Let's add the necessary methods to support this usage.

```
//FILE: ea_pcix_trans.h
struct pcix_trans {
  int devnum;
  int addr;
  int attr1;
  int attr2;
  int cmd;
  int data[8];
  bool done;
  // Required by sc_signal<> and sc_fifo<>
  ea_pcix_trans& operator= (const ea_pcix_trans& rhs)
  {
    devnum = rhs.devnum;   addr   = rhs.addr;
    attr1  = rhs.attr1;    attr2  = rhs.attr2;
    cmnd   = rhs.cmnd;     done   = rhs.done;
    for (unsigned i=0;i!=8;i++) data[i]=rhs.data[i];
    return *this;
  }
  // Required by sc_signal<>
  bool operator== (const ea_pcix_trans& rhs) const {
    return (
      devnum ==rhs.devnum && addr   == rhs.addr   &&
      attr1  ==rhs.attr1  && attr2  == rhs.attr2  &&
      cmnd   ==rhs.cmnd   && done   == rhs.done   &&
      data[0]==rhs.data[0]&& data[1]== rhs.data[1]&&
      data[2]==rhs.data[2]&& data[3]== rhs.data[3]&&
      data[4]==rhs.data[4]&& data[5]== rhs.data[5]&&
      data[6]==rhs.data[6]&& data[7]== rhs.data[7]);
  }
};
// Required functions by SystemC
ostream& operator<<(ostream& file,
                    const ea_pcix_trans& trans);

void sc_trace(sc_trace_file*& tf,
              const ea_pcix_trans& trans,
              sc_string nm);
```

*Figure 13-4.* Example of SystemC User Data Type

We provide example implementations of the latter two methods here:

```cpp
//FILE: ea_pcix_trans.cpp
#include <systemc.h>
#include "ea_pcix_trans.h"
ostream& operator<<(ostream& os,
                    const ea_pcix_trans& trans)
{
  os << "{" << endl << "   "
     << "cmnd: " << trans.cmnd   << ", "
     << "attr1:" << trans.attr1  << ", "
     ...
     << "done:"  << (trans.done?"true":"false")
     << endl << "}";
  return os;
} // end
// trace function, only required if actually used
void sc_trace(sc_trace_file*& tf,
              const ea_pcix_trans& trans,
              sc_string nm)
{
  sc_trace(tf, trans.devnum,  nm + ".devnum");
  sc_trace(tf, trans.addr,    nm + ".addr");
  ...
  sc_trace(tf, trans.data[7], nm + ".data[7]");
  sc_trace(tf, trans.done,    nm + ".done");
} // end trace
```

*Figure 13-5.* Example of SystemC User Data Type Implementation

It should be noted that the **sc_trace** method is only necessary if used; however, best practices suggest that you should always provide this method. Observe that this method is always expressed in terms of other traces that are already defined (e.g., the built-in ones).

In some cases, it may be difficult to determine an appropriate representation. For example, **char\*** or **string** have no real equivalent. In these cases, you may either convert to an unsigned fixed-bit-width vector (e.g., **sc_bit**), or omit it completely. However, remember that converting these representations is for ease of debug and doing so is usually of much more value than you might think. The same may be said of appropriate representation for **ostream**.

You may also wish to implement **ifstream** or **ofstream** to support verification needs.

As you can see, the added support is really quite minimal, and it is only required for custom data types.

## 13.4 The Heartbeat, a Custom Hierarchical Channel

Hierarchical channels are interesting, because they're really hybrid modules. Technically, a hierarchical channel must inherit from **sc_channel**; however, **sc_channel** is really just a #**define** for **sc_module**. Hierarchical channels must also inherit from an interface to allow them to be used with **sc_port**.

Why would you define a hierarchical channel? One use of hierarchical channels is to model complex buses such as PCI, AMBA™, or HyperTransport™. Another common use of hierarchical channels, adaptors and transactors will be discussed in the next section.

To keep things simple, we'll model a simple clock or heartbeat. This clock will differ from the standard hardware concept that typically uses a Boolean signal. Instead, our heartbeat channel will issue a simple event. This usage would correspond to the **posedge_event** used by so many hardware designs.

Because it's simple, the heartbeat is much more efficient simulation-wise than a Boolean signal. Here is the header for our simple interface:

```
struct ea_heartbeat_if: public sc_interface {
  virtual const sc_event& default_event() const = 0;
  virtual const sc_event& posedge_event() const = 0;
};
```

*Figure 13-6.* Example of Hierarchical Interface Header

It's no different than a primitive channel interface. Notice that we use method names congruent with **sc_signal**. This convention will simplify design refinement. The careful design of interfaces is key to reducing work that is done later.

Let's look at the corresponding channel header, which inherits from **sc_channel** instead of **sc_prim_channel** and has a process, **SC_METHOD**.

```cpp
#include "ea_heartbeat_if.h"
struct ea_heartbeat
: public sc_channel, public ea_heartbeat_if {
  SC_HAS_PROCESS(ea_heartbeat);
  // Constructor (only one shown)
  explicit ea_heartbeat(sc_module_name nm,
                        sc_time _period)
  : sc_channel(nm)
  , m_period(_period)
  {
    SC_METHOD(heartbeat_method);
    sensitive << m_heartbeat;
  }
  // User methods
  const sc_event& default_event() const
  { return m_heartbeat; }
  const sc_event& posedge_event() const
  { return m_heartbeat; }
  void heartbeat_method(); // Process
private:
  sc_event m_heartbeat; // *The* event
  sc_time  m_period;    // Time between events
  // Copy constructor so compiler won't create one
  ea_heartbeat( const ea_heartbeat& );
};
```

*Figure 13-7.* Example of Hierarchical Channel Header

Let's see how it's implemented:

```
#include <systemc.h>
#include "ea_heartbeat.h"

void ea_heartbeat::heartbeat_method(void) {
  m_heartbeat.notify(m_period);
}
```

*Figure 13-8.* Example of Hierarchical Channel Interface Header

In the next chapter, we'll see the built-in SystemC clock, which has more flexibility at the expense of performance.

## 13.5 The Adaptor, a Custom Primitive Channel

Also known in some circles as transactors, adaptors are a type of channel specialized to translate between modules with different interfaces. Adaptors are used when moving between different abstractions. For example, an adaptor is commonly used between a test bench that models communications at the transaction level (i.e., TLM), and an RTL implementation that models communications at the pin-accurate level. Transaction-level communications might have methods that transfer an entire packet of information (e.g., a PCI-X transaction). Pin-accurate level communications use Boolean signals with handshakes, clocks, and detailed timing.

To make it easy to understand, we're going to investigate two adaptors. In this section, we'll see a simple primitive channel that uses the evaluate-update mechanism. In the following section, we'll investigate a hierarchical channel. For many of the simpler communications, an adaptor needs nothing more than some member functions and a handshake to exchange data. This often meets the requirements of a primitive channel. Actually, many of the simpler adaptors could go either way, since they don't require an evaluate-update mechanism.

**Adaptor - Primitive Channel**

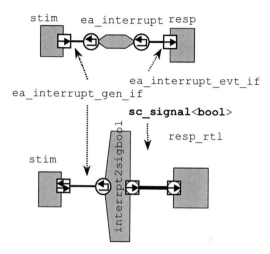

*Figure 13-9.* Before and After Adaptation

We will now discuss an example design. The example design includes a stimulus (stim) and a response (resp) that are connected via an **ea_interrupt** channel described in an earlier section. We now would like to replace resp with a refined RTL version, resp_rtl, that requires an **sc_signal<bool>** channel interface. The before and after example design is graphically shown in *Figure 13-9*.

Here is the adaptor's header:

```
#include "ea_interrupt_gen_if.h"
struct interrupt2sigbool
: public sc_prim_channel
, public ea_interrupt_gen_if
, public sc_signal_in_if<bool>
{
  // Constructors
  explicit interrupt2sigbool()
  : sc_prim_channel(
      sc_gen_unique_name("interrupt2sigbool")) {}
  explicit interrupt2sigbool(sc_module_name nm)
  : sc_prim_channel(nm) {}
  // Methods for ea_interrupt_gen_if
  void notify() {
    m_delay = SC_ZERO_TIME; request_update(); }
  void notify(sc_time t) {
    m_delay = t; request_update(); }
  // Methods for sc_signal_in_if<bool>
  const sc_event& value_changed_event() const
  { return m_interrupt; }
  const sc_event& posedge_event() const
  { return value_changed_event(); }
  const sc_event& negedge_event() const
  { return value_changed_event(); }
  const sc_event& default_event() const
  { return value_changed_event(); }
  // true if last delta cycle was active
  const bool& read() const {
    m_val = event(); return m_val;
  }
  // get a reference to current value (for tracing)
  const bool& get_data_ref() const { return read(); }
  // was there a value changed event?
  bool event() const {
    return (simcontext()->delta_count()==m_delta+1);
  }
// continued next figure...
```

*Figure 13-10.* Example of Primitive Adaptor Channel Header (1 of 2)

```
  bool posedge() const { return event(); }
  bool negedge() const { return event(); }
  const sc_signal_bool_deval& delayed() const {
    const sc_signal_in_if<bool>* iface = this;
    return RCAST<const sc_signal_bool_deval&>
                    ( *iface );
  }
protected:
  // every update is a change
  void update() {
    m_interrupt.notify(m_delay);
    m_delta = simcontext()->delta_count();
  }
private:
  sc_event m_interrupt;
  mutable bool m_val;
  sc_time  m_delay;
  uint64   m_delta;    // delta of last event
  // Copy constructor so compiler won't create one
  interrupt2sigbool( const interrupt2sigbool& );
};
```

*Figure 13-11.* Example of Primitive Adaptor Channel Header (2 of 2)

The first thing to notice is all the methods. Most of these are forced upon us because we are inheriting from the **sc_signal_in_if**<**bool**> class. Fortunately, most of them may be expressed in terms of others for this particular adaptor. Another way to handle excess methods is to provide stubbed error messages with the assumption that nobody will use them.[34]

The second feature of interest is the manner in which evaluate-update is handled. In the **notify**() methods, we update the delay and make a **request_update**() call to the scheduling kernel. When the delta-cycle occurs, the kernel will call our **update**() function that issues the appropriately delayed notification.

For the most part, this adaptor was simple. The hard part was obtaining a list of all the routines that needed to be implemented as a result of the interface. Listing the routines is accomplished easily enough by simply

---

[34] If you use this approach, we strongly suggest the error messages be implemented with the error reporting mechanism discussed in Chapter 14, Advanced Topics and be classified as SC_FATAL.

examining the interface definition in the Open SystemC Initiative library
source code.

A third feature to note is the use of the call to
`simcontext()->delta_count()`, which is not documented in the LRM.
While we don't encourage this type of freedom, it is sometimes necessary.
Note that the LRM and SystemC have not finalized on the standard
specification at the time of this book's first edition. We believe a few details
like this need to be added to the LRM. The main point of this example is the
adaptor.

Finally, for those not completely up on their C++, a comment on the
`mutable bool`. The keyword `mutable` means changeable even if `const`.
This keyword is used for situations like this. The `read()` method is defined
as `const` in the `sc_signal_in_if<bool>` interface. So we have to
implement it, and the implementation must satisfy the `const` directive (i.e.,
it may not change any internal state). It also is required to return a reference
(`&`). We are using the member function `event()` to obtain the value, which
is not a reference. So, we create a member data `m_val` which can be used by
reference. In order to store the return value of `event()` we have to change
the unchangeable; hence, `mutable`. We are not violating the spirit of the
`const`.

## 13.6 The Transactor, a Custom Hierarchical Channel

When a more complex communications interface is encountered, such as
one that requires processes, hierarchy, or ports, then a hierarchical channel
solution is required. Such is the case with the following processor interface
problem. On one side, we have a peripheral, an 8K x 16 memory that is
normally controlled by a processor. On the other side, we have a test bench
that needs to use simple transaction calls to verify the functionality of the
memory. This adaptor is often referred to as a transactor because it allows
the test bench to convert transactions into pin-level stimulus.

Graphically, here are the elements of the design:

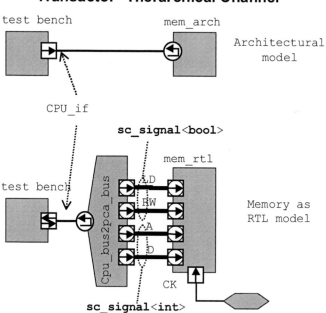

## Transactor - Hierarchical Channel

*Figure 13-12.* Test Bench Adaptation Using Hierarchical Channels

There are actually potentially two hierarchical channels in this picture. The architectural model of the memory is a module implementing an interface, in this case the CPU_if.

Let's take a look at the CPU interface:

```
struct CPU_if: public sc_interface {
  virtual void write(unsigned long addr, long data)=0
  virtual long read(unsigned long  addr)=0;
};
```

*Figure 13-13.* Example of Simple CPU interface

The corresponding memory implementation is a trivial channel:

```
//FILE: mem_arch.h
#include "CPU_if.h"
struct mem: public sc_channel, CPU_if {
  // Constructors & destructor
  explicit mem(sc_module_name nm,
               unsigned long ba, unsigned sz)
  : sc_channel(nm), m_base(ba), m_size(sz)
  { m_mem = new long[m_size]; }
  ~mem() { delete [] m_mem; }
  // Interface implementations
  virtual void write(unsigned long addr, long data) {
    if (m_start <= addr && addr < m_base+m_size) {
      m_mem[addr-m_base] = data;
    }
  }
  virtual long read(unsigned long  addr) {
    if (m_start <= addr && addr < m_base+m_size) {
      return m_mem[addr-m_base];
    } else {
      cout << "ERROR:"<<name()<<"@"<<sc_time_stamp()
           << ": Illegal address: " << addr << endl;
      sc_stop(); return 0;
    }
  }
private:
  unsigned long m_base;
  unsigned      m_size;
  long          m_mem[];
  mem(const mem&); // Disable
};
```

*Figure 13-14.* Example of Hierarchical Channel Implementation

Now, suppose we have the following timing diagram for the pin-cycle accurate interface:

## CPU PCA Timing

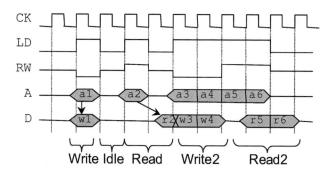

*Figure 13-15.* CPU Pin-Cycle Accurate Timing

Notice that write transactions take place in a single clock cycle; whereas, read has a one-cycle delay for the first read in a burst. Also, this interface assumes a bi-directional data bus. Address and read/write have a non-asserted state. We'll allow this to be a don't care situation for this design.

Here is the transactor's header:

```
#include "CPU_if.h"
#include "ea_heartbeat_if.h"
SC_MODULE(cpu2pca), CPU_if {
  // Ports
  sc_port<ea_heartbeat_if> ck; // clock
  sc_out<bool>             ld; // load/execute command
  sc_out<bool>             rw; // read high/write low
  sc_out<unsigned long>    a;  // address
  sc_inout_rv<32>          d;  // data
  // Constructor
  SC_CTOR(cpu2pca) {}
  // Interface implementations
  void write(unsigned long addr, long data);
  long read(unsigned long addr);
private:
  cpu2pca(const cpu2pca&); // Disable
};
```

*Figure 13-16.* Example of Hierarchical Transactor Channel Header

Clearly, with the preceding example, the basics of a module are present, but the addition of inheriting from `CPU_if` simply adds some methods to be implemented, namely **read**() and **write**().

An interesting point to ponder with channels (especially adaptors and transactors) is the issue of member function collisions. What if two or more interfaces that need to be implemented have identically-named member functions with identical argument types? There are two solutions. Prior to SystemC version 2.1, this situation required renaming or otherwise modifying the interfaces. This is an ugly situation, but cannot be avoided. With the advent of **sc_export**<> in SystemC 2.1, it is possible to isolate each interface to an **sc_export**<>.

Here is the implementation code for the transactor:

```
#include <systemc.h>
#include "cpu2pca.h"
enum operation {WRITE=false, READ=true};
void cpu2pca::write(unsigned long addr, long data) {
  wait(ck->posedge_event());
  ld->write(true);
  rw->write(WRITE);
  a->write(addr);
  d->write(data);
  wait(ck->posedge_event());
  ld->write(false);
}
long cpu2pca::read(unsigned long addr) {
  wait(ck->posedge_event());
  ld->write(true);
  rw->write(READ);
  a->write(addr);
  d->write(FLOAT);
  wait(ck->posedge_event());
  ld->write(false);
  return d->read().to_long();
}
```

*Figure 13-17.* Example of Hierarchical Transactor Implementation

The code for an adaptor/transactor can be very straightforward. For more complex applications, such as a PCI-X bus interface, the design of an adaptor may be more complex.

Transactors and adaptors are very common in SystemC designs, because they allow high-level abstractions to interface with lower-level implementations. Sometimes these hybrids are used as part of a design refinement process. At other times, they merely aid the development of verification environments. There are no fixed rules about which levels have to use them.

## 13.7 Exercises

For the following exercises, use the samples provided in www.EklecticAlly/Book/.

**Exercise 13.1**: Examine, compile, and run the `interrupt` example. Write a specialized port for this channel to support the method **pos** ().

**Exercise 13.2**: Examine, compile, and run the `pcix` example. Could this process of converting a **struct** to work with an **sc_signal** be automated?

**Exercise 13.3**: Examine, compile, and run the `heartbeat` example. Extend this channel to include a programmable time offset.

**Exercise 13.4**: Examine, compile, and run the `adapt` example. Notice the commented out code from the adaptation of `resp` to `resp_rtl`.

**Exercise 13.5:** Examine, compile, and run the `hier_chan` example. Examine the efficiency of the calls. Extend the design to allow back-to-back reads and writes in a cycle efficient manner.

# Chapter 14

## ADVANCED TOPICS
*Clocks, Clocked Threads, Programmable Hierarchy, Signal Tracing, & Dynamic Processes*

Congratulations for keeping up to this point. This section begins with a basic discussion of clocks and then quickly accelerates to a discussion of **SC_CTHREAD**s, programmable hierarchy, waveform tracing, and a very advanced discussion of dynamic threads and **SC_FORK/SC_JOIN**.

If you are able to follow this section, you are ready to take on the world. However, if you become discouraged, come back and reread the chapter after gaining a little SystemC coding experience.

## 14.1 sc_clock, Predefined Processes

Clocks represent a common hardware behavior, a repetitive Boolean value. If you are a hardware designer, it is likely you've been concerned about the late discussion of this topic. This topic is delayed for a reason.

Clocks add many events, and much resulting simulation activity is required to update those events. Consequently, clocks can slow simulations significantly. Additionally, quite a lot of hardware can be modeled adequately without clocks. If you need to delay a certain number of clock cycles, it is much more efficient to execute a wait for the appropriate delay than to count clocks as illustrated in *Figure 14-1*.

```
wait(N*t_PERIOD) // one event -> FAST!
-OR-
for(i=1;i<=N;i++) // creates many events -> slow
  wait(clk->posedge_event())
```

*Figure 14-1.* Comparing Wait Statements to Clock Statements

More importantly, many designs can be modeled without any delays. It all depends on information to be derived from the model at a particular stage of a project.

A clock can be easily modeled with SystemC. Indeed, we have already seen an example of a clock modeled with just an event, namely the heartbeat example. More commonly, clocks are modeled with an **sc_signal**<**bool**> and the associated event.

Clocks are so common that SystemC provides a built-in hierarchical channel known as an **sc_clock**. Clocks are commonly used when modeling low-level hardware where clocked logic design currently dominates.

```
sc_clock name("name",period
             [,duty_cycle=0.5,first=0,rising=true]);
```

*Figure 14-2.* Syntax of sc_clock

Notice the optional items indicated by their defaults.

Some caveats apply to **sc_clock**. First, if declared within a module, **sc_clock** must be declared and initialized prior to its use. Second, if you want to communicate a clock as an output to the module, you must provide a process sensitive to the clock that assigns the output signal port. An alternative approach available with SystemC version 2.1 is to use an **sc_export**<**sc_signal_in_if**<**bool**> >.

For example:

```
SC_MODULE(clock_gen) {
  sc_port<sc_signal_out_if<bool> >  clk1_p;
  sc_export<sc_signal_in_if<bool> > clk2_p;//SysC2.1
  sc_clock clk1;
  sc_clock clk2;
  SC_CTOR(clock_gen)
  : clk1("clk1",4,SC_NS)
  , clk2("clk2",6,SC_NS)
  {
    SC_METHOD(clk1_method);
    sensitive << clk1;
    clk2_p(clk2);
  }
  void clk1_method() {
    clk1_p->write(clk1);
  }
};
```

*Figure 14-3.* Example of sc_clock

Consider that the first method used for **clk1** involves additional activity, which inevitably slows the simulation. This method also entails more code.

## 14.2 Clocked Threads, the SC_CTHREAD

SystemC has two basic types of processes: the **SC_THREAD** and the **SC_METHOD**. A variation on the **SC_THREAD**, the clocked **SC_CTHREAD**, is popular for behavioral synthesis tools. This popularity is partly because synthesized logic tools currently produce fully synchronous code, and it is partly because the **SC_CTHREAD** provides some new facilities to simplify coding.

```
SC_CTOR(module_name) {
  SC_CTHREAD(NAME_cthread, clock_name.edge());
}
```

*Figure 14-4.* Syntax of SC_CTHREAD

Some of the simpler facilities provided by this new process are a new form of **wait()** and a level-sensitive wait, called **wait_until()**. The syntaxes are:

```
wait(N); // delay N clock edges
wait_until(delay_expr); // until expr true @ clock
```

*Figure 14-5.* Syntax of Clocked Waits

In addition, the delay expression, **delay_expr**, must be expressed using delayed signals. In other words, all arguments for **wait_until()** must be of the form *signal*.**delayed()**. The **delayed()** method is a special method that provides the value at the end of a delta-cycle.

Neither of these is extremely interesting. They are correspondingly almost equivalent to the following **SC_THREAD** code assuming the thread is statically sensitive to a clock edge:

```
for(i=0;i!=N;i++) wait();//similar as wait(N)
do wait() while(!expr);// sames as
                       // wait_until(dexpr)
```

*Figure 14-6.* Example of Code Equivalent for Clocked Thread

Although the preceding is functionally equivalent to the statements referenced in the comments, the **SC_CTHREAD** implementations are faster. This improved speed is because **SC_CTHREAD** implementation does not have to actually resume and suspend at each intermediate **wait()**. Notice that **wait_until()** is not really level sensitive since it only tests at the clock edge defined for the process.

Of greater interest, **SC_CTHREAD** provides the concept of watched signals, which effectively changes the behavior of **wait()**. When a watched signal activates, execution jumps to a new area upon return from **wait()** rather than proceeding to the next statement. SystemC has two forms of watching: global and local.

The simplest form of watching is global. The effect of globally watched signal activation is to restart the clocked thread at the beginning.

The syntax is simple and follows:

```
SC_CTOR(module_name) {
  SC_CTHREAD(NAME_cthread);
  watching(signal.delayed()=true);
}
```

*Figure 14-7.* Syntax of Global Watching

By contrast, a more controllable form of watching involves using four macros, **W_BEGIN**, **W_DO**, **W_ESCAPE**, and **W_END**. These macros are always used as a set in the order indicated, and they define three regions in the code being watched.

In the first region, the signals being watched are declared. In the second region, the code being watched is coded. In the last region, the code to handle the occurrence of watched signal activation is provided. Here is a synopsis of the syntax:

```
W_BEGIN
  watching(delay_expr);
W_DO
  Code_being_watched
W_ESCAPE
  Code_handling_escape_condition
W_END
```

*Figure 14-8.* Syntax of Local Watching

This syntax is equivalent to the following code:

```
#define WAIT(A) wait(A); if (expr) throw(W_ESCAPE);
try {
  Code_being_watched // use WAIT() instead of wait()
} catch(W_ESCAPE) {
  Code_handling_escape_condition
}
```

*Figure 14-9.* Example of Code Equivalent to Local Watching

The advantage of the syntax is coding simplification. Also, these extensions allow for nesting and combination with the global watching. Doing this with ordinary code quickly becomes onerous.

An example should help in understanding. This code models a simple processor (no caching) used in the design of a collision-avoidance system for a futuristic automobile.

```
#include "processor.h"
SC_HAS_PROCESS(processor);
processor::processor(sc_module_name nm) //Constructor
: sc_module(nm)
{
  // Process registration
  SC_CTHREAD(processor_cthread,clock_p.pos());
  watching(reset_p.delayed() == false);
}//endconstructor }}}
void processor::processor_cthread() { //{{{
  // Initialization
  pc = RESET_ADDR;
  wait();
  for(;;) {
    W_BEGIN
      watching(abort_p.delayed() == true);
    W_DO
      read_instr();
      switch(opcode) {
        case STORE_ACC:
          bus_p->write(operand1,acc);
          break;
        case INCR:
          acc++;
          result = (acc != 0);
          break;
        ...
    W_ESCAPE
      cout << sc_time_stamp()
           << " WARN: Aborting" << endl;
    W_END
  }//endforever
}//endcthread
```

*Figure 14-10.* Example Code Using Clocked Threads and Watching

We'll note one last point. Unlike **SC_THREAD**, which upon exiting never runs again, **SC_CTHREAD** restarts on the next clock edge. If you want to stop an **SC_CTHREAD**, then call the **halt()** method.

There has been some discussion of deprecating **SC_CTHREAD**. However, **SC_THREAD** functionality may need to be augmented by the extra mechanisms of watching and the resulting simplified syntax before eliminating this feature.

## 14.3 Programmable Hierarchy

Programmable elaboration is an aspect of SystemC that may be obvious to some but not to others. The code that performs elaboration (i.e., instantiates modules and channels) is simply executable C++ code. This means that it is possible to use standard C++ constructs such as **if-then-else**, **switch**, **for**, and **while** loops.

Thus, it is conceivable to have simulations that configure themselves. In some cases, this configurability is a matter of convenience. For instance, configurability is appropriate for a large regular structure. In other cases, configurability may be a way to test various aspects of the design. Let's look at a couple of examples.

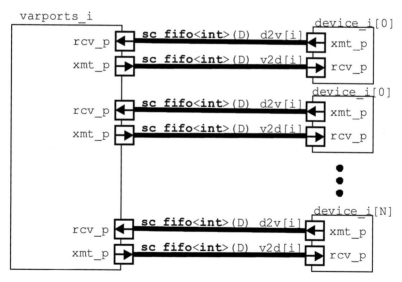

*Figure14-11.* Design with 1-*N* Ports

First, we consider a design that supports a variable number of devices attached externally. Take for example, an Ethernet or USB port. The specification diagram looks something like the previous figure.

To test this design, the verification team would like a single executable that can be configured when run to handle 0 to 16 devices with varying FIFO depths. Here is the supporting code:

```
#include <systemc.h>
#include "varports.h"
#include "device.h"
int main(int argc, char* argv[]) {
  /* Figure out N from command-line */
  ...
  varports*      varports_i;
  device*        device_i[N]; //N previously set to 16
  sc_fifo<int>* v2d[N];
  sc_fifo<int>* d2v[N];
  varports_i = new varports(...init parameters...);
  // nDevices set through command line or equivalent
  for (unsigned i=0;i!=nDevices;i++) {
    sc_string nm; // for unique instance names
    // Create instances
    nm = sc_string::to_string("device_name_i[%d]",i);
    device_i[i] = new device(nm.c_str());
    nm = sc_string::to_string("v2d[%d]",i);
    v2d[i] = new sc_fifo<int>(nm.c_str(),fifo_depth);
    nm = sc_string::to_string("d2v[%d]",i);
    d2v[i] = new sc_fifo<int>(nm.c_str(),fifo_depth);
    // Connect devices to varports using channels
    device_i[i]->rcv_p(*v2d[i]);
    device_i[i]->xmt_p(*d2v[i]);
    varports_i->rcv_p(*d2v[i]);
    varports_i->xmt_p(*v2d[i]);
  }//endfor
  ...
```

*Figure 14-12.* Example of Configurable Code with 1-*N* Ports

The preceding example uses arrays of pointers to both the instances and the channels connecting them. We could have dynamically set the array size; however, it would not save enough resources to justify the complexity and effort.

Our second example recognizes the importance of configuration management. A design may start out with a TLM and eventually be refined to RTL. It is desirable to be able to run simulations that easily select portions of the design to run at TLM or RTL levels. TLM portions will simulate quickly; RTL portions will represent something closer to the final implementation and will simulate more slowly. This configurability lets the verification engineer keep simulations running quickly, and he or she can focus on finding problems in a particular area.

Configurability may be achieved by using conditional code (e.g., **if-else**) around the areas of interest. For example, consider the hierarchical channel design of the previous chapter (hier_chan example). Suppose we package both the architectural model and the behavioral model within a wrapper that lets us configure the design at run time.

We can read the configuration instance names into an STL **map**<>. An example of the wrapper code is shown in Figure 14-13. The code shown defaults to an architectural implementation, mem_arch. Both an RTL and bsyn configuration are supported; although, the selection of an RTL version only produces a warning message.

```
SC_MODULE (mem)  {
  sc_export<CPU_if>         CPU_p;
  mem_arch*                 mem_arch_i;
  mem_bsyn*                 mem_bsyn_i;
  cpu2pca*                  cpu2pca_i;
  ea_heartbeat*             clock;
  sc_signal<bool>           ld;
  sc_signal<bool>           rw;
  sc_signal<unsigned long>  a;
  sc_signal_rv<32>          d;
  SC_HAS_PROCESS (mem);
  explicit mem(sc_module_name nm,
               unsigned long ba, unsigned sz)
  : sc_channel (nm)
  {
    if (cfg[name()] == "rtl") {
      cout << "WARN: " cfg[name()]
           << " not yet supported " << endl;
    }
    if (cfg[name()] == "bsyn") {
      clock = new ea_heartbeat("clock",
                               sc_time(10,SC_NS));
      mem_bsyn_i = new mem_bsyn("mem_bsyn_i",ba,sz);
      mem_bsyn_i->ld(ld);   mem_bsyn_i->rw(rw);
      mem_bsyn_i->a(a);     mem_bsyn_i->d(d);
      mem_bsyn_i->ck(*clock);
      cpu2pca_i = new cpu2pca("cpu2pca_i");
      cpu2pca_i->ld(ld);   cpu2pca_i->rw(rw);
      cpu2pca_i->a(a);     cpu2pca_i->d(d);
      cpu2pca_i->ck(*clock);
      CPU_p(*cpu2pca_i);
    } else {
      mem_arch_i = new mem_arch("mem_arch_i",ba,sz);
      CPU_p(*mem_arch_i);
    }//endif
  }
};
```

*Figure 14-13.* Example of Configurable Code to Manage Levels

## 14.4 Debugging and Signal Tracing

Until this point, we have assumed the use of standard C++ debugging techniques such as in-line print statements or using a source code debugger such as gdb. Hardware designers are familiar with using waveform-viewing tools that display values graphically.

While SystemC does not have a built-in graphic viewer, it can copy data values to a file in a format compatible with most waveform viewing utilities. The format is known as VCD or Value Change Dump format. It is a simple text format.

Obtaining VCD files involves three steps. First, open the VCD file. Next, select the signals to be traced. These two steps occur during elaboration. Running the simulation (i.e., calling **sc_start**()) will automatically write the selected data to the dumpfile. Finally, close the trace-file. Here is the syntax presented in sequence:

```
sc_trace_file* tracefile;
tracefile = sc_create_vcd_trace_file(tracefile_name);
if (!tracefile) cout <<"There was an error."<<endl;
...
sc_trace(tracefile,signal_name,"signal_name");
...
sc_start(); // data is collected
...
sc_close_vcd_trace_file(tracefile);
```

*Figure 14-14.* Syntax to Capture Waveforms

It is required that the signal names being traced are defined before calling **sc_trace**. Also, it is possible to use hierarchical notation to access signals in sub-modules. It is possible to trace ordinary C++ data values, and ports as well. The trace filename should not include the filename extension since the **sc_create_vcd_trace_file** automatically does this. Notice the error checking of the file creation using **operator**!().

Here is a simple coding example:

```
//FILE: wave.h
SC_MODULE(wave) {
  sc_signal<bool> brake;
  sc_trace_file*  tracefile;

  ...

  double temperature;
};
```

```
//FILE: wave.cpp
wave::wave(sc_module_name nm) //Constructor
: sc_module(nm) {

  ...

  tracefile = sc_create_vcd_trace_file("wave");
  sc_trace(tracefile,brake,"brake");
  sc_trace(tracefile,temperature,"temperature");
}//endconstructor
wave::~wave() {
    sc_close_vcd_trace_file(tracefile);
    cout << "Created wave.vcd" << endl;
}
```

*Figure 14-15.* Example of Simple Waveform Capture

Notice the use of a destructor to close the file. This is the safest way to ensure the file will be closed. If additional modules are instantiated in the example above, they would need to include appropriate **sc_trace** syntax within their constructors.

Another moderately complex example of signal tracing may be found in the tracing example from the book website.

Here is some sample output viewed with the opensource `gtkwave`[35] viewer:

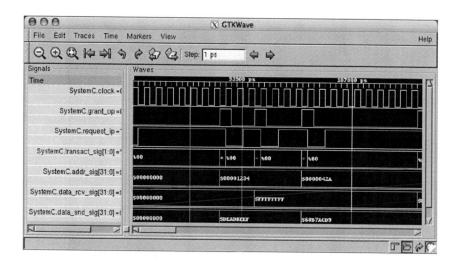

*Figure 14-16.* Sample Waveform Display From gtkwave

In addition to VCD files, SystemC supports WIF (Waveform Interchange Format) simply by using the **sc_create_wif_trace_file** call. Some EDA vendors may also provide their own custom formats and support.

## 14.5 Dynamic Processes

Thus far, all the process types discussed have been static. In other words, once the elaboration phase completes, all **SC_THREAD**, **SC_METHOD**, and **SC_CTHREAD** processes have been established. SystemC 2.1 introduces the concept of dynamically spawned processes. This concept is not new. In the examples that came with SystemC 2.0.1, there was a fork-join example; however, it was not considered part of the official SystemC release. This feature was added with SystemC 2.1.

Dynamic processes are important for several reasons. At the top of the list is the ability to perform temporal checks such as those supported by PSL Sugar, Vera, and other verification languages.

For instance, consider a bus protocol with split transactions and timing requirements. Once a request is issued, it is important to track the

---

[35] Available from http://www.cs.man.ac.uk/apt/tools/gtkwave/index.html

completion of that transaction from a verification point of view. Since transactions may be split, each transaction will require a separate thread to monitor. Without dynamic process support, it would be necessary to pre-allocate a number of statically defined processes to accommodate the maximum number of possible outstanding requests.

Let's look at the syntax and requirements that enable dynamic processes.

First, to enable dynamic processes, it is necessary to use a pre-processor macro prior to the invocation of the SystemC header file.

Here's one way to do this:

```
#define SC_INCLUDE_DYNAMIC_PROCESSES
#include <systemc.h>
```

*Figure 14-17.* Syntax to Enable Dynamic Threads

Other mechanisms involve the C++ compilation tools. For example, GNU gcc has a **-D** option (e.g., **-DSC_INCLUDE_DYNAMIC_PROCESSES**).

Next, functions need to be declared for use as processes. Functions may be either normal functions or methods (i.e., member functions of a class). The dynamic facilities of SystemC version 2.1 allow for either **SC_THREAD** or **SC_METHOD** style processes. Unlike static processes, dynamic processes may have up to eight arguments and a return value. The return value will be provided via a reference variable. For example, consider the following declarations:

```
void inject();//ordinary func w/ no args or return
int count_changes(sc_signal<int>& sig);//ordinary
                                       //function
bool TestChan::Track(sc_signal<packet>& pkt);//meth
void TestChan::Errors(int maxwarn, int maxerr);//meth
```

*Figure 14-18.* Example Functions Used as Dynamic Processes

Having declared and defined functions or methods to be used as processes, only registering them with the kernel remains. You can register the dynamic processes within an **SC_THREAD** or with restrictions within an **SC_METHOD**.

The basic syntax is as follows:

```
sc_process_handle hname = sc_spawn(//ordinary
                                   //function
  /*void*/sc_bind(&funcName, ARGS...),//no return value
  processName,
  spawnOptions
);
sc_process_handle hname = sc_spawn(//member function
  /*void*/sc_bind(&methName, object, ARGS...)//no
                                             //return
  processName,
  spawnOptions
);
```

*Figure 14-19.* Syntax to Register Dynamic Processes with Void Return

```
sc_process_handle hname = sc_spawn(//ordinary
                                   //function
  &returnVar, sc_bind(&funcName, ARGS...)
  processName,
  spawnOptions
);
sc_process_handle hname = sc_spawn(//member function
  &returnVar, sc_bind(&methodName, object, ARGS ...)
  processName,
  spawnOptions
);
```

*Figure 14-20.* Syntax to Register Dynamic Processes with Return Values

Note in the preceding that *object* is a reference to the calling module, and normally we just use the C++ keyword **this**, which refers to the calling object itself.

By default, arguments are passed by value. To pass by reference or by constant reference, a special syntax is required:

```
sc_ref(var)   // reference
sc_cref(var)  // constant reference
```

*Figure 14-21.* Syntax to Pass Process Arguments by Reference

The *processName* and *spawnOptions* are optional; however, if *spawnOptions* are used, then a *processName* is mandatory. All processes should have unique names. Fortunately, uniqueness of a process name includes the hierarchical instance as a prefix (i.e., **name**()). If a process spawns a process, then its name is used to prefix the spawned process name.

Spawn options are determined by creating an **sc_spawn_option** object and then invoking one of several methods that set the options.

Here is the syntax:

```
sc_spawn_option objname;
objname.set_stack_size(value);
objname.set_method();//register as SC_METHOD
objname.dont_initialize();
objname.set_sensitivity(event_ptr);
objname.set_sensitivity(port_ptr);
objname.set_sensitivity(interface_ptr);
objname.set_sensitivity(event_finder_ptr);
```

*Figure 14-22.* Syntax to Set Spawn Options

One last comment before we look at an example. The method **sc_get_cur_process_handle**() may be used by the spawned process to reference the calling object. In particular, it may be useful to access **name**().

That's a lot of syntax. Fortunately, you don't need to use all of it. Let's take a look at an example of the simplest case. This example is an **SC_THREAD** that contains no parameters and returns no result. In other words, it looks like an **SC_THREAD** that just happens to be dynamically spawned. We highlight the important points.

```
#define SC_INCLUDE_DYNAMIC_PROCESSES
#include <systemc.h>
...
void spawned_thread() {// This will be spawned
  cout << "INFO: spawned_thread "
       << sc_get_curr_process_handle()->name()
       << " @ " << sc_time_stamp() << endl;
  wait(10,SC_NS);
  cout << "INFO: Exiting" << endl;
}
void simple_spawn::main_thread() {
  wait(15,SC_NS);
  sc_spawn(sc_bind(&spawned_thread));
  cout << "INFO: main_thread " << name()
       << " @ " << sc_time_stamp() << endl;
  wait(15,SC_NS);
  cout << "INFO: main_thread stopping "
       << " @ " << sc_time_stamp() << endl;
}
```

*Figure 14-23.* Example of a Simple Thread Spawn

If you keep a handle on the spawned process, then it is also possible to await the termination of the process via the **sc_process_handle::wait**() method. For example:

```
sc_process_handle h =
             sc_spawn(sc_bind(&spawned_thread));
// Do some work
h.wait(); // wait for the spawned thread to return
```

*Figure 14-24.* Example of Waiting on a Spawned Process

Be careful not to wait on an **SC_METHOD** process; currently; there is no way to terminate an **SC_METHOD**. Extensions to address this shortcoming are expected in SystemC version 2.2.

An interesting observation about **sc_spawn** is that it may also be used within the constructor, and it may be used with the same member function multiple times as long as the process name is unique. This capability also means there is now a way to pass arguments to **SC_THREAD** and **SC_METHOD** as long as you are willing to use the longer syntax.

A dangerous aspect of spawned threads relates to the return value. If you pass a function or method that returns a value, it is critical that the referenced return location remain valid throughout the lifetime of the process. The result will be written without respect to whether the location is valid upon exit, possibly resulting in a really nasty bug.

The creation and management of dynamic processes is not for the faint of heart. On the other hand, learning to manage dynamic processes has great rewards. One of the simpler ways to manage dynamic processes is discussed in the next section on fork/join.

## 14.6 SC_FORK/SC_JOIN

Another use of dynamic threads is dynamic test configuration. This feature is exemplified with a verification strategy sometimes used by Verilog suites using fork/join. Although this technique does not let you create new modules or channels dynamically (because processes may choose to stimulate ports differently on the fly), you can reconfigure tests. Let's see how this might be done.

Consider the DUT in the following figure.

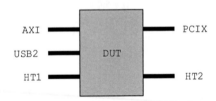

*Figure 14-25.* High-Level Model of a Design to be Tested

For each interface (AXI, USB2, etc.), an independent process can be created either to send or receive information likely to be generated in a real system.

Using these processes and fork/join, a high-level test might look as follows (Note: Syntax will be explained shortly):

```
DataStream d1, d2;
SC_FORK
  sc_spawn(sc_bind(&AXI_xmt,this,sc_ref(d1)),"p1"),
  sc_spawn(sc_bind(&PCI_rcv,this,sc_ref(d1)),"p2"),
  sc_spawn(sc_bind(&HT1_xtm,this,sc_ref(d2)),"p3"),
  sc_spawn(sc_bind(&PCIX_rcv,this,sc_ref(d2)),"p4")
SC_JOIN
```

*Figure 14-26.* Example of fork/join Application

Here is the syntax for SystemC fork/join:

```
SC_FORK
  COMMA_SEPARATED_LIST_OF_SPAWNS
SC_JOIN
```

*Figure 14-27.* Syntax for fork/join

Let's look at an example that involves a number of syntax elements discussed thus far. First, let's inspect the header for this module.

```
//FILE: Fork.h
SC_MODULE(Fork) {
  ...
  sc_fifo<double> wheel_lf, wheel_rt;
  SC_HAS_PROCESS(Fork);
  Fork(sc_module_name nm);// Constructor
  // Declare a few processes to be used with
  // fork/join
  void fork_thread();
  bool road_thread(sc_fifo<double>& which);
};
```

*Figure 14-28.* Example Header for fork/join Example

Notice that we pass a FIFO channel by reference so that `road_thread` can possibly access the FIFO channel. Passing ports or signals by reference would be a natural extension of this idea.

Now, let's inspect the code:

```cpp
//FILE: Fork.cpp
#define SC_INCLUDE_DYNAMIC_PROCESSES
#include <systemc.h>
#include "Fork.h"
...
Fork::Fork(sc_module_name nm) //{{{
: sc_module(nm)
{
  SC_THREAD(fork_thread);
  ...
}
void Fork::fork_thread() { //{{{
  bool lf_up, rt_up; // use for return values
  SC_FORK
    sc_spawn(&lf_up,
             sc_bind(&Fork::road_thread, this,
                     sc_ref(wheel_lf)
             ),
             "lf" // process name
    ) /*endspawn*/,
    sc_spawn(&rt_up,
             sc_bind(&Fork::road_thread, this,
                     sc_ref(wheel_rt)
             ),
             "rt" // process name
    ) /*endspawn*/
  SC_JOIN
}//end Fork:fork_thread
bool Fork::road_thread(sc_fifo<double>& which) { //
  // Do some work
  return (road > 0.0);
}//end Fork::road_thread
```

*Figure14-29*. Example of fork/join

The full example may be found in the downloaded examples as `Fork`. This example also illustrates the use of **sc_spawn** instead of **SC_THREAD**. Using a capitalized word `Fork` was done to avoid collision with the Unix system call `fork`, which has nothing to do with SystemC's **SC_FORK**. Recall that SystemC is a cooperative multi-tasking system. Please don't confuse Unix's fork facilities with these concepts.

## 14.7 Error and Message Reporting

Reporting messages is an important art, and many a project has created utilities to standardize this reporting within the project. Messages have classifications including informational, warning, error, and fatal. Additionally, messages usually apply to a variety of areas and need to be isolated to their source to aid debugging. For simulations, it is also important to identify the time that a message occurs. Because simulations provide a tremendous amount of output data, it is important that messages be standardized and easy to identify.

SystemC's version 2.1 adds an error reporting system that greatly simplifies this task. Throughout our examples thus far, you have seen a stylized format of error management. In this short section, we will examine a subset of the error-reporting facilities in SystemC version 2.1. For more information, you are referred to the SystemC version 2.1 release notes and the example documentation that accompanies the release.

We need a few definitions first. Every message is associated with an identifying name. This labeling is used to keep messages from different parts of the design properly identified. A message identifier is simply a character string:

```
char* MSGID = "UNIQUE_STRING";
```

*Figure 14-30.* Syntax of Message Identifier

Next, all messages need to be classified. SystemC version 2.1 has the following classifications:

| | |
|---|---|
| **SC_INFO** | – informational only |
| **SC_WARNING** | – possible problem |
| **SC_ERROR** | – problem identified probably serious |
| **SC_FATAL** | – extremely serious problem probably ending simulation |

*Figure 14-31.* Error Classifications

For each classification, a variety of actions may be taken. For the most part, defaults are sufficient. Possible actions include the following actions taken from the SystemC version 2.1 example documentation:

**SC_UNSPECIFIED** – Take the action specified by a configuration rule of a lower precedence.
**SC_DO_NOTHING** – Don't take any actions for the report, the action will be ignored, if other actions are given.
**SC_THROW** – Throw a C++ exception (**sc_exception**) that represents the report. The method **sc_exception::get_report**() can be used to access the report instance later.
**SC_LOG** – Print the report into the report log, typically a file on disk. The actual behavior is defined by the report handler function.
**SC_DISPLAY** – Display the report to the screen, typically by writing it into the standard output channel using **std::cout.**
**SC_INTERRUPT** – Interrupt simulation if simulation is not being run in batch mode. Actual behavior is implementation defined, the default configuration calls **sc_interrupt_here**(...) debugging hook and has no further side effects.
**SC_CACHE_REPORT** – Save a copy of the report. The report could be read later using **sc_report_handler::get_cached_report**(). The reports saved by different processes do not overwrite each other.
**SC_STOP** – Call **sc_stop**(). See **sc_stop**() manual for further detail.
**SC_ABORT** – The action requests the report handler to call **abort**() .

*Figure 14-32.* Error Actions

SystemC has a large class of setup that may be specified for message reporting. For basic designs, the following syntax should suffice:

```
sc_report_handler::set_log_file_name("filename");
sc_report_handler::stop_after(SC_ERROR, MAXERRORS);
sc_report_handler::set_actions(MSGID, CLASS, ACTIONS);
```

*Figure 14-33.* Syntax for Basic Message Setup

The following code, named `report`, illustrates the basics of message handling:

```
char* sim_name = "mysim";
char* sim_vers = "$Header$";
int sc_main(int argc, char* argv[]) {
  sc_report_handler::set_log_file_name("run.log");
  sc_report_handler::stop_after(SC_ERROR, 100);
  sc_report_handler::set_actions(
    sim_name,
    SC_INFO,
    SC_CACHE_REPORT|SC_LOG
  );
  SC_REPORT_INFO(sim_name,sim_vers);
  .../* Body of main */
  sc_start();
  sc_report* rp =
    sc_report_handler::get_cached_report();
  if ( rp ) {
    cout << rp->get_msg() << endl;
    cout << sim_name << " FAILED" << endl;
    return 1;
  } else {
    cout << sim_name << " PASSED" << endl;
    return 0;
  }
}
```

*Figure 14-34.* Example of main.cpp with SystemC Error Reporting

```
extern char* sim_name;
void mymod::some_thread() {
  wait(2,SC_NS);
  SC_REPORT_INFO(sim_name,"Sample info");
  SC_REPORT_WARNING(sim_name,"Sample warning");
  SC_REPORT_ERROR(sim_name,"Sample error");
  SC_REPORT_FATAL(sim_name,"Sample fatal");
}
```

*Figure 14-35.* Example of Reporting in a Module

Here is a sample of the output:

```
0 s: Info: mysim: $Header: /.../mysim/main.cpp,v 1.2... $
2 ns: Info: mysim: Sample info
2 ns: Warning: mysim: Sample warning
In file: mymod.cpp:21
In process: mymod_i.some_thread @ 2 ns
2 ns: Error: mysim: Sample error
In file: mymod.cpp:22
In process: mymod _i.some_thread @ 2 ns
...
```

*Figure 14-36.* Example of Output Messages

You can enhance the output by using a syntax-highlighting editor and setting up a coloring scheme for log files.

## 14.8 Other Libraries: SCV, ArchC, and Boost

Beyond the core of SystemC, several libraries are available for the serious SystemC user to explore. These include:

- The SystemC Verification library, the SCV, has an extensive set of features useful for verification. The original set was donated by Cadence Design Systems; their www.testbuilder.net website is the best source of information.

- The ArchC architecture description language is an open-source architecture description language used to describe processors. Several models are already available. ArchC was designed at the Computer Systems Laboratory (LSC) of the Institute of Computing of the University of Campinas (IC-UNICAMP). See www.archc.org for more information.

- The Boost website provides free peer-reviewed portable C++ source libraries. The emphasis is on libraries that work well with the C++ Standard Library. See www.boost.org for more information.

## 14.9 Exercises

For the following exercises, use the samples provided at www.EklecticAlly.com/Book/.

**Exercise 14.1**: Examine, compile, and run the `clock_gen` example.

**Exercise 14.2**: Examine, compile, and run the `processor` example.

**Exercise 14.3**: Examine, compile, and run the `varports` example.

**Exercise 14.4**: Examine, compile, and run the `manage` example. Can you think of a simpler way to manage different implementations that leverages C++?

**Exercise 14.5**: Examine, compile, and run the `cruisin` example.

**Exercise 14.6**: Examine, compile, and run the `wave` example. View the VCD data using a waveform viewer. Obtain `gtkwave` from http://www.cs.man.ac.uk/apt/tools/gtkwave/index.html if necessary.

**Exercise 14.7**: Examine, compile, and run the `tracing` example.

**Exercise 14.8**: Examine, compile, and run the `simple_spawn` example. Modify it to spawn an **SC_METHOD** that is statically sensitive to a signal of your choice with **dont_initialize**().

**Exercise 14.9**: Examine, compile, and run the `Fork` example. Can you nest fork/join? Explain how to do this.

**Exercise 14.10**: Examine, compile, and run the `report` example. Apply these concepts to an earlier example.

# Chapter 15

## ODDS & ENDS
*Performance, Gotchas, and Tidbits*

This chapter wraps up with a few simple observations on using SystemC to its greatest advantage. The authors provide hints about ways to keep simulation performance high and provide observations about the modeling language in general. This chapter contains no exercises. We leave application to your individual creativity.

## 15.1 Determinants in Simulation Performance

We sometimes hear comments from folks such as, "We tried SystemC, but our simulations were slower than Verilog." Such comments betray a common misconception. SystemC is not a faster simulator. The Open SystemC Initiative reference version of the SystemC simulator has several opportunities for optimization, and there are EDA vendors hoping to capitalize on that situation. More importantly, simulation performance is not so much about the simulator as it is the way the system is modeled.

KEY POINT:    SystemC simulation speed is linked directly to the use of higher levels of modeling using un-timed and transaction-level concepts.

For all simulators (e.g., SPICE, Verilog, VHDL, or SystemC), there are fundamental tasks that must be performed: moving data, updating event queues, keeping track of time, etc. Any simulator simulating detailed pin-level activity and timing information will provide a certain level of performance. Almost all simulators for a given class of detail will perform within a factor of two or so.

No so long ago, cycle-based simulators were all the rage due to their advertised speed. Problems arose when designers discovered that these simulators didn't provide the same level of accuracy as their event-driven counterparts, but that was exactly the reason they ran faster!

That said, RTL simulates at RTL speeds. Certainly, there are simulators that do RTL better than others, but they still have the limitation of keeping track of all the same details. A good optimizer may improve performance by finding commonalities, but the improvement will be bounded.

GUIDELINE:   To improve simulation performance, reduce details and model at higher levels of abstraction whenever possible.

It is possible to obtain dramatic speed improvements by keeping as much of the system as possible at very high levels of abstraction, and only using details where absolutely required.

Part of the problem lies with understanding what a given simulation is supposed to accomplish. Ask yourself, "What is this simulation model supposed to answer?" For example, early in the design process the architect may wish to know if a new algorithm even works. At this level, timing and pins are not really interesting. A simple executable that takes input data and produces output for analysis is all that is required. Timing should not be a part of this model. Sequential execution is probably sufficient.

Another set of questions might be, "Have all the parts been connected? Have we defined paths for all the information required to perform the system functions?" These questions may be answered by creating a module for every component and using a simple transaction-level model to interconnect the pieces. Cycle accuracy should not be a concern at this point in the design.

### 15.1.1   Saving Time and Clocks

How can you live without time or clocks[36]? This is really quite simple. For instance, suppose you need to model time to determine performance. Rather than coding a wait for $N$ rising edges, it is much more efficient to simply delay by $N*$`clock_period`.

Another common technique occurs when using handshakes. If you need to wait for a signal, then simply wait on the signal directly. The hardware may do sampling at clock edges, but that wastes time.

---

[36] We ask this question from an electronic system design perspective, not from a philosophical perspective.

If you really need to synchronize to the clock, then do both, as follows:

```
wait(acknowledge->posedge_event());
if (!clock->event()) wait(clock->posedge_event());
```

*Figure 15-1.* Synchronized wait for a Signal

Perhaps you need to transfer information from one port to another in the design. Even though you know the result will be delayed through a FIFO over multiple clocks, there is no need to create a FIFO. Just read it from the input, delay, and write it to the output.

```
In->read(packet);
wait(50*CLOCK_PERIOD);
Out->write(packet);
```

*Figure 15-2.* Example of FIFO Elimination

Events are also a powerful way of communicating information. If you don't really need to test the value of a signal but are only interested in the change, it is more efficient to use an event than a `sc_signal<bool>`. Earlier in the book, we illustrated some primitive channels to do just this (e.g., in the `heartbeat` example).

Another overlooked issue is using too much resolution. Does your clock really need to oscillate at 100 MHz? Perhaps it would suffice to use a higher-level clock. Do you really need to measure picoseconds, or are nanoseconds or even milliseconds sufficient?

## 15.1.2   Moving Large Amounts of Data

So, the model is efficiently using time, but it still appears to be simulating too slowly. Perhaps you are attempting to move too much data. Do you really need to move the data or do you just need to record the fact that data was moved and that an appropriate amount of time has passed?

For example, perhaps you could model the movement of chunk of data as follows instead of moving the actual data:

```
struct payload {
  unsigned long byte_count;
  unsigned value; // a single unique value
};
```

*Figure 15-3.* struct for Payload

Now, you will need to modify the read/write routines in the channels to do something like this:

```
void Bus<payload>::write(unsigned addr, payload data)
{
  wait(data.byte_count*t_BYTE_DELAY);
  // transfer the data
}
```

*Figure 15-4.* Bus Write with Payload

Perhaps, you need to transfer the data, but how much data do you really need to test the problem at hand? For instance, if dealing with video graphics, would a small 64 x 48 pixel buffer suffice to test an algorithm, rather than a full 640 x 480 or larger frame?

Perhaps, you need to transfer a large block of data across the bus, but can you model it using smart pointers instead? In other words, manage the chunk of simulator memory with a pointer. We recommend you use a Boost.org smart pointer, or the equivalent, to avoid problems with memory leaks or corruption.

Thus, you might have:

```
struct payload {
  unsigned long byte_count;
  smart_ptr<int> pValues;
  payload(unsigned long bc)
  : byte_count(bc)
  {
    pValues = new int[bc];
  }
};
```

*Figure 15-5.* Smart Pointer with Payload

Now, you are simply passing around a pointer and only manipulating the data when it really needs to be manipulated.

Do you really need to fully populate a memory, or would a sparse memory model suffice? The SystemC Verification library contains a very nice sparse memory model that is very easy to use.

### 15.1.3 Too Many Channels

Another interesting area for SystemC designers to watch is channels. Every channel interaction involves at least two calls (producer and consumer), two events, and possibly two copy operations. Hierarchical **sc_port** to **sc_port** connections are very efficient because they simply pass a pointer to the target channel at elaboration time. With the advent of SystemC version 2.1, you will also find **sc_export** to **sc_export** hierarchical connections are similarly efficient. If you find yourself writing a process that simply copies one port to another, consider the possibility of rearchitecting the connectivity.

### 15.1.4 Effects of Over Specification

Often designers tend to think in terms of the final implementation rather than the general problem being designed. This approach sometimes results in over specification. For instance, a behavior may be specified as a finite state machine (FSM), when the real issue is simply a handshake or data transfer. Be careful when presented with myriads of detail to abstract the real needs of the design.

### 15.1.5  Keep it Native

Keeping data native has already been discussed under data types earlier in the book, but this topic bears repeating. Data types are an abused subject. Does the model at hand really need to specify 17 bits, or would a simple **int** suffice? Native C++ data types will simulate many times faster than their SystemC hardware-specific counterparts. Similarly, what do you gain using **sc_logic**? Is the unknown value relevant to the current level of modeling? Once again, the issue is to model only those items that will affect the results of the simulation.

### 15.1.6  C++ Compiler Optimizations

Depending on the stability of your model, you may want to consider looking at optimizing your use of the C++ compiler. Many times, default `make` scripts assume that the developer wants maximum debug visibility and the compiler obliges with additional visibility that may affect simulation performance.

When looking for maximum performance, make sure that your SystemC library and your system design are compiled without a debug option. Additionally, some compilers have switches that perform additional runtime optimizations at the expense of increased compile time. If you plan to run extensive simulation with the same model, it may pay to wade through the documentation for your compiler.

Another example, ensuring that **#ifndef** is on the first line of a header file improves performance for some compilers.

## 15.2  Features of the SystemC Landscape

Because SystemC is a C++ class library rather than a truly independent language, SystemC has some aspects that seem to annoy its users (particularly experienced designers from an RTL background). This section simply notes these aspects. Keep in mind that part of the power of SystemC is the fact that it is C++, and therefore, it is extremely compatible with application software.

### 15.2.1  Things You Wish Would Just Go Away

For the novice, just getting a design to compile can be a challenge. This section lists some of the most common problems. All of them relate directly to C++.

Syntax errors in **#include** files often are reported as errors in the including implementation (i.e., .cpp file). The most common error is forgetting to put the trailing semicolon on a **SC_MODULE**, which is really a **class**.

The use of semicolons in C++ may seem odd at times. The **class** and **struct** require a closing semicolon.

```
class myclass {
   // Body            Required
};                    semicolon
```

*Figure 15-6.* C++ Class Requires Semicolon

On the other hand, function definitions and code blocks do not require a semicolon.

```
void myfunction {
   // Body
}                     No semicolon
```

*Figure 15-7.* C++ Function Does Not Use Semicolon

Similarly, **SC_FORK/SC_JOIN** have the odd convention of using commas. This punctuation is used because they are really just fancy macros.

```
SC_FORK                  No braces
   sc_spawn (...),
   sc_spawn (...),       Commas except
   sc_spawn (...)        for last one
SC_JOIN
                         No semicolon
```

*Figure15-8.* C++ Fork/Join Idiosyncrasy

SystemC relies heavily on templates. The templates have the annoying space between the greater than brackets.

```
sc_port<sc_signal_in_if<int> > data_ip;
```

Required space

*Figure 15-9.* C++ Template Idiosyncrasy

Inside the basic syntax of a module, **sensitive** and **dont_initialize** methods must be tied to the immediately preceding **SC_THREAD**, **SC_METHOD** or **SC_CTHREAD** registration. This tying is usually a lot easier to deal with if you indent the code slightly relative to the registration. For example:

```
SC_CTOR(SomeModule) {
  SC_METHOD(sync_method);
    sensitive << clock;
    dont_initialize();
  SC_METHOD(monitor_method);
    sensitive << rqst << ack;
    dont_initialize();
  SC_THREAD(compute_thread);
}
```

Indented relative to registration

*Figure 15-10.* Example of Using Indents to Highlight Registrations

**dont_initialize** brings up another issue common to **SC_METHOD**. Unless you specify otherwise, all processes are executed once at initialization despite the appearance of static sensitivity unless **dont_initialize** is used. For some, this can be confusing at first. Try to remember that all simulation processes are run during initialization unless **dont_initialize** is applied.

## 15.2.2 Feature Solutions

What is the best way to address these idiosyncrasies besides just learning them? We highly recommend obtaining language-sensitive text editors with color highlighting, and we recommend obtaining lint tools designed specifically for SystemC. The authors' favorite text editor is vim in graphical mode, also known as gvim. You can obtain a copy of vim from www.vim.org for almost any platform.

Other users are quite successful using emacs (graphical of course) or nedit. All three of these editors have environments available for download that support SystemC. You can obtain these from our website.

There are a few C++ lint tools and at least one lint tool focused on SystemC that is commercially available[37]. Some EDA tools have built-in lint-like checkers. Your mileage will vary, and we highly recommend a careful evaluation before committing to any of these tools.

### 15.2.3 Conventions and Coding Style

Coding styles are a well-known issue. Probably one of the best books written on this subject for hardware design is the *Reuse Methodology Manual for System-on-a-Chip Designs* by Michael Keating and Pierre Bricaud. Most of the concepts presented there have direct application to SystemC. Let's just touch on a few.

A name is a name, right? Wrong! Names of classes, variables, functions and other matters are a critical part of making your code readable and understandable. If you have been observant, you will notice we've inserted various naming conventions specific to SystemC in the examples. For instance, processes always have a suffix of `_thread`, `_method` or `cthread`. This convention is used because **wait**() results in a runtime error when used with **SC_METHOD**, and visa versa for **next_trigger**().

Similarly, we adopted a convention when addressing ports and probably you should do likewise for using anything **sc_signal** or otherwise supporting the evaluate-update paradigm.

Thus, it should come as no surprise that the authors are developing a recommended coding style for SystemC and that we are planning to support it with commercially available lint tool rule-sets.

---

[37] Actis Design www.actisdesign.com.

## 15.3 Next Steps

If you have read this far, you are probably considering adopting SystemC for an upcoming project. Or, perhaps you have already started, and you are looking for help moving forward. This section provides some ideas.

### 15.3.1    Guidelines for Adopting SystemC

In the fall of 2003, the authors presented a paper on the subject of language adoption, "How to Really Mess Up Your Project Using a New Language" at the Synopsys User's Group in Boston, MA. We included a number of key points, which we provide for your consideration.

1. Don't do it alone - obtain management support.
2. Doing things the same way will produce the same results regardless of the language.
3. Look at the big picture, the product or system - not the small tasks.
4. Don't skimp on training - obtain good formal training.
5. Obtain mentoring.
6. Adopting the new paradigm is necessary to gain the advantages of a new language.
7. Specifications should use the appropriate level of abstraction for the new paradigm.
8. Put coding discipline in place quickly with coding guidelines, lint tools, and reviews.
9. Choose templates approved by seasoned experts in the new language.
10. Start automation and environment simply and cleanly.
11. Evaluate EDA tools for the big picture.
12. Insist on well-documented and supported tools in all areas including tools version and configuration.
13. Apply the technology to a pilot project that focuses on the big picture.

There are a number of companies supporting SystemC methodologies. A quick visit to the OSCI website www.systemc.org can provide a starting point. Or, visit our website www.EklecticAlly.com for our view.

### 15.3.2 Resources for Learning More

For the readers who would like more information on SystemC extensions, we recommend the following resources for further study.

*Table 15-1.* SystemC Resources

| | Type | Details |
|---|---|---|
| 1 | Website | Starting point for SystemC. Retrieved March 2004 from: http://www.systemc.org/. This site has several great papers and white papers as well as email forums for getting help or discussing SystemC. |
| 2 | Website | Starting point for SystemC Verification Library. Retrieved March 2004 from: http://www.testbuilder.net/ |
| 3 | Website | The European SystemC Users Group web site. This site has additional quality papers and additional news and activities. Retrieved March 2004 from: http://www-ti.informatik.uni-tuebingen.de/~systemc/ |
| 4 | Website | The web site for the recently formed North American SystemC Users Group. Focused on SystemC activities in North America. Retrieved March 2004 from: http://www.nascug.org |
| 5 | Website | References for SystemPerl HDL tools. Retrieved March 2004 from: http://www.veripool.com/ |

For the readers still gasping for help with C++, here are some additional recommendations for further study.

*Table 15-2.* C++ Resources

|   | Type | Details |
|---|------|---------|
| 1 | Book | Koenig, A., Moo, B., *Accelerated C++*. Boston: Addison-Wesley, 2000. A highly recommended textbook for learning to speak C++ natively. |
| 2 | Book | Bjarne Stroustrup. The C++ Programming Language. Florham Park, New Jersey:Addison_Wesley, 2000. Probably the best C++ reference and is written by the creator of C++. |
| 3 | Book | Kyle Loudon. *C++ Pocket Reference*. Sebastopol, California: O'Reilly & Associates, Inc., 2003. A convenient and reasonably organized quick reference. Good for those who are not yet C++ experts. |
| 4 | Book | Nicolai M Josuttis. *The C++ Standard Library: A Tutorial and Reference*. Indianapolis, Indiana: Addison-Wesley, 1999. A complete reference and good tutorial for the STL, a very useful library for modeling. |
| 5 | Website | Stroustrup, B. Definitive reference for C++ by the author of C++. Retrieved March 2004 from: http://www.research.att.com/~bs/C++.html |
| 6 | Tool | van Heesch, D. Documentation system for C++ code. Retrieved March 2004 from: http://www.stack.nl/~dimitri/doxygen/index.html |
| 7 | Website | References for C++ programming. Retrieved March 2004 from: http://www.cplusplus.com/ |
| 8 | Book | Henricson, M., Nyquist. E., *Industrial Strength C++*. Upper Saddle River, New Jersey: Prentice Hall, 1996. (Online Version: http://www.elho.net/dev/doc/industrial-strength.pdf) |
| 9 | Web book | A free online book. Retrieved March 2004 from: http://www.mindview.net/Books/TICPP/ThinkingInCPP2e.html |

| 10 | Article | Hoff, T. C++ Coding Standard. Retrieved March 2004 from: http://oopweb.com/CPP/Documents/CodeStandard/ VolumeFrames.html. |
|----|---------|---|
| 11 | Article | Baldwin, J. 1992. An Abbreviated C++ Code Inspection Checklist. Retrieved March 2004 from: http://www.chris-lott.org/resources/ cstyle/Baldwin-inspect.pdf |
| 12 | Book | Pressman, R., *Software Engineering: A Practitioner's Approach*. McGraw-Hill, 2001. (Online Version: http://www.rspa.com/about/sepa.html) |
| 13 | Web class | Free C++ class based on an inexpensive tool. Retrieved March 2004 from: http://www.codeWarriorU.com/ |
| 14 | Web class | Free C++ class. Retrieved March 2004 from: http://www.free-ed.net/fr03/lfc/030203/120/ |

We hope you'll be ready for our next book when we introduce topics such as the SystemC Verification library, and we go deeper into a discussion of SystemC design methodologies and design styles. We also expect to provide updates as SystemC version 2.2 that is just now appearing on the horizon.

# Acknowledgments

- Our inspiration was provided by:
  - Mike Baird, President of Willamette HDL, who provided basic knowledge
  - Wes Campbell of Eklectic Ally Inc. who brought us together
- Our reviewers provided feedback that helped keep us on track:
  - Chris Donovan
  - Ronald Goodstein, First Shot Logic Simulation and Design
  - Mark Johnson, LogicMeister Consulting
  - Rob Keist
  - Chris Macionski, Bright Eyes Consulting
  - Nasib Naser, Synopsys Inc.
  - Suhas Pai, Qualcomm Incorporated
- Our Graphic Artist
  - Felix Castillo
- Our Technical Editors helped us say what we meant to say:
  - Kyle Smith, Smith Editing
  - Richard Whitfield

Most important of all, we acknowledge our wives, Pamela Black and Carol Donovan. These wonderful women (despite misgivings about two wild-eyed engineers) supported us cheerfully as we spent many hours researching, typing, discussing, and talking to ourselves while pacing around the house as we struggled to write this book over the past year.

We also acknowledge our parents who gave us the foundation for both our family and professional life.

# List of Figures

**Reader's Notes**

# Index

3516580

Made in the USA